THE EVERYTHING® GUIDE TO PRE-ALGEBRA

Dear Reader,

I've always felt that "pre-algebra" was a bit of a misnomer. Sure, some of the information learned in a course on pre-algebra is indeed pre-algebra, but plenty of it is just regular old algebra. There's nothing "pre" about it.

Switching from arithmetic to algebra is a big shift in terms of how you think and how to practice math. Some people find that shift easy, but most people find it very, very tricky. Whether you find it easy or hard says nothing about your ability to succeed at math. Let me say that again: If the new algebra skills you're learning don't come easily to you, that says nothing about whether or not you're "good" at math. Everything in math is learnable with practice.

I hope that this book helps you to work through the core concepts of pre-algebra and provides some insight into those concepts.

Jane Cassie

Welcome to the EVERYTHING® Series!

These handy, accessible books give you all you need to tackle a difficult project, gain a new hobby, comprehend a fascinating topic, prepare for an exam, or even brush up on something you learned back in school but have since forgotten.

You can choose to read an Everything® book from cover to cover or just pick out the information you want from our four useful boxes: e-questions, e-facts, e-alerts, and e-ssentials.

We give you everything you need to know on the subject, but throw in a lot of fun stuff along the way, too.

We now have more than 400 Everything® books in print, spanning such wide-ranging categories as weddings, pregnancy, cooking, music instruction, foreign language, crafts, pets, New Age, and so much more. When you're done reading them all, you can finally say you know Everything®!

QUESTION

Answers to common questions

FACT

Important snippets of information

ALERT

Urgent warnings

ESSENTIAL

Quick handy tips

PUBLISHER Karen Cooper

MANAGING EDITOR, EVERYTHING® SERIES Lisa Laing

COPY CHIEF Casey Ebert

ASSISTANT PRODUCTION EDITOR Alex Guarco

ACQUISITIONS EDITOR Pamela Wissman

DEVELOPMENT EDITOR Eileen Mullan

EVERYTHING® SERIES COVER DESIGNER Erin Alexander

THE EVERYTHING® GUIDE TO PRE-ALGEBRA

A helpful practice guide through the pre-algebra basics—in plain English!

Jane Cassie, EdM

Avon, Massachusetts

An Everything® Series Book.

Published by Adams Media, a division of F+W Media, Inc.
57 Littlefield Street, Avon, MA 02322 U.S.A.
www.adamsmedia.com

ISBN 10: 1-4405-6640-2
ISBN 13: 978-1-4405-6640-0
eISBN 10: 1-4405-6641-0
eISBN 13: 978-1-4405-6641-7

Printed in the United States of America.

10 9 8 7 6 5 4 3 2 1

Library of Congress Cataloging-in-Publication Data

Cassie, Jane, author.
The everything guide to pre-algebra / Jane Cassie, EdM.
pages cm. -- (An everything series book)
ISBN-13: 978-1-4405-6640-0 (pbk.)
ISBN-10: 1-4405-6640-2 (pbk.)
ISBN-13: 978-1-4405-6641-7 (electronic)
ISBN-10: 1-4405-6641-0 (electronic)
1. Mathematics. I. Title. II. Title: Pre-algebra.
QA107.2.C37 2013
512.9--dc23
2013021389

*This book is available at quantity discounts for bulk purchases.
For information, please call 1-800-289-0963.*

Contents

Acknowledgments

So much of our love or fear of math is shaped by our teachers. I want to thank Mr. Walkins, my high school math teacher, for teaching me that anything is learnable with practice and questions. You have shaped the paths of many lives.

Thanks of course go to my editor, Pam Wissman, for believing in this project, and to Addam Stine, my constant source of support.

Top 10 Reasons Why Pre-Algebra Is Worth Mastering

1. Everything in the study of math is built on pre-algebra. With a strong foundation there, the rest of the building will go much more smoothly.
2. Unlike a lot of what you learn in school, pre-algebra skills do have a direct application to almost any job or career you might want.
3. Someone's going to give you a grade, and grades matter. (What? You thought it was *just* for the love of math? You know you're in it for the grade!)
4. Pre-algebra ties together a lot of concepts you've studied previously.
5. Math is kind of amazing. It's actually really fun to see the different ways the concepts come together when you give it a chance.
6. The feeling of success that comes when you *really* understand a concept and know why it's true, instead of just memorizing that it is true, is truly great.
7. Once you know something, no one can take it away from you. Pre-algebra is a chance to learn a set of tools you can use for the rest of your education.
8. You *can* learn anything in this book with the right explanation and enough practice. And if you can master something that will help you, you should!
9. Understanding math gives you huge leverage when you decide what to study in high school, college, and beyond. Fields in science, business, economics, healthcare, sports, communication, and art all use these skills.
10. Honestly, almost no one learns algebra just for the love of it; someone's making you learn it. So you may as well get the most out of it!

Introduction: Why Bother Learning Algebra?

ALGEBRA IS OFTEN A struggle for many students. That's because it is an entirely new way of thinking about math for them. Up until pre-algebra, most math classes are based on computation. They're based on numbers only, and focused on learning what you're allowed to do with those numbers. Then algebra comes along, and a big shift happens. All of a sudden, instead of just numbers, you have variables, which are letters that hold the place of numbers. And that means you have to think about the rules of numbers and the way they work, instead of just doing the computations you've been practicing for many years. That's tough!

At first, learning algebra may be frustrating, because it might seem like it's pointless. Rest assured, though, that it is far from pointless. Even though you probably won't do a lot of algebra in your daily life as an adult (unless you teach algebra for a living), understanding algebra is a useful skill you'll use again and again in a variety of ways.

First and foremost, it helps you understand how to think about numbers in a new way, and to solve problems. That's how your brain learns to process all kinds of new information, which is something everyone does every day. It's like doing pushups if you're a football player. Sure, pushups aren't really a part of a football player's professional performance, but doing them builds a muscle that is vital to playing the sport properly.

Don't overlook the fact that success in math is a requirement for many fields of study. Whether it should be or shouldn't be doesn't matter—it is. Does that mean you can't be a success without getting good at algebra? Absolutely not. But it does mean that if you want to go to school to be a doctor, scientist, veterinarian, accountant, lawyer (yes, lawyer), businessperson, finance professional, teacher, or just about anything else that requires a college degree, you'll need to be able to do algebra.

The thing about algebra is that it builds on itself, so if you miss one concept, it can start to get really confusing really quickly. I hope this book will help you get unstuck in places where you are confused.

CHAPTER 1

Study Strategies

Let's face it: becoming good at math takes work. Not only do you have to study; you have to study hard. But studying hard isn't enough. If you're studying the wrong way, all of your work is not going to pay off. After all, practice only makes things permanent, but practicing the right way makes them perfect. This means that if you're practicing incorrectly, you're just cementing bad habits into place. That's actually even worse than doing nothing! And you certainly shouldn't waste your time doing work that isn't helping. Before you start learning all the details of pre-algebra math, it is important to review *how* to study pre-algebra. This chapter will give you some ideas as to how you can effectively study so that you can use this book in the way that's most helpful to you. Your goal should be to pick up a few pointers, so that when the quiz or test or final comes along, you'll be able to show what you've learned!

Don't Call Yourself "Bad at Math"

All students encounter subjects that come naturally to them and concepts that they find a little bit harder to understand. It can be really frustrating to work on something when you feel like the people around you don't have to work quite as hard. But the fact of the matter is that you're reading this book because you have to learn math. Whether it's to get ready for high school or to pass a college course, learning pre-algebra is a step you have to complete to get where you want to go.

Most people who believe they are "bad at math" usually have one or two concepts that they didn't completely understand somewhere within their math education, whether those steps were in second grade or in middle school. When that happens, they have a weak foundation. And when they start trying to build other stuff on top of that foundation, it doesn't go very well. That doesn't mean they are bad at math. It just means they have to find the area or areas they don't really understand and spend a little time fixing the foundation in that area.

If you "don't like math," it might be time to switch your mindset. Instead of focusing on liking or disliking math, focus on liking the feeling of understanding something. When the goal is to understand a concept instead of to enjoy that concept, it's easier to achieve.

Don't ever say you "can't do math." Can you split a bag of candy in half? Can you figure out how much money two things will cost when you know the price of each one? Can you calculate how much money you'll earn this month if you earn $5 every day? Then you can do math. Math on paper (and in class, on quizzes, and on tests) is just a code that represents what happens with numbers in the real world. Learn the code, one piece at a time, and you'll be able to do it.

Ask for Help

This is one of the most important actions you can take, because it's something that students don't often think to do. Asking for help doesn't mean that you can't do it yourself. It doesn't mean you're stupid. It doesn't mean you don't know what you're doing.

What it means is that you're a good student. Good students ask for help, wherever they can get it. That's how you learn. The people who learn the most in life are the ones who ask questions.

Whether you're totally lost, or you just have one little question, get an answer. Math builds on itself, and if there is a hole in your understanding, your foundation for future topics is going to be weak. Ask your classmates. Ask your older brother or sister. Ask your parent or tutor. Ask in a forum online. And, of course, ask your teacher! Don't just ask the day before the test—ask as soon as the problem comes up.

There are so many great resources for learning and practicing pre-algebra online. If the explanation in your textbook or class doesn't make sense to you, try doing a Google or YouTube search for the topic at hand. Sometimes a slight change in the wording of an explanation is enough to make something click!

Explain It to Someone Else

Sometimes you hear your teacher explain a concept in class, and it makes perfect sense. Then you get home to do your homework, and you just can't figure out how to get to the answer. Other times, it doesn't make sense even during class, but everyone else seems to get it.

If you think you don't understand a concept, try explaining it to someone else. Maybe you'll be able to explain it, and then you'll see that you really do understand it. If not, you'll be able to clarify what it is that you're having trouble with, because it will be where you get stuck during the explanation. Explaining is a way to make sure you understand the reasoning behind each step. When you understand the reasoning, you'll really understand what's going on. That's the road to success.

This is really hard to do, but it works. It's much harder to try to explain something than to have it explained to you, because it makes you vulnerable. But don't be afraid to make a mistake: this is an excellent way to learn. Try to explain to someone who already knows the material so that he or she can tell you if you're making a mistake and help you when you get stuck. Give it a try!

Take It One Bite at a Time

Some math problems involve using ten or twenty different skills. It looks like one problem, but it has a whole series of intricate steps and skills that have to be done correctly to get the right answer!

If you are having trouble, break the problem down one step at a time to see if you can catch where your mistake is occurring. Maybe you actually understand the concept, but you're making an arithmetic mistake. Maybe you are doing all the arithmetic right, but you don't understand exactly what the question is asking you to find. If you can figure out which step you're having trouble with, you can get a better explanation, re-read your textbook, or do some more practice problems of the same type.

Don't try to go too fast. Everyone learns at his or her own speed, and if you don't understand a concept yet, you haven't really learned it. If you feel like class is going too fast for you, you should try to schedule some extra practice time each week. That will slow down the speed at which you have to learn, which will allow you to go at your own pace.

Cement Good Habits

Doing something correctly once is not enough to say that you know it. It's proof that you can do it, but it's just a start to really learning. Once you've learned something, you know how to do it when it shows up again.

That means that you have to practice the steps you take to solve a problem enough times that they are consistent and comfortable in the future. The best way to practice math is to practice in short, frequent study sessions. When you try to cram while learning a skill, you don't get the most out of the problems. If you do thirty of the same problem type in a row, you are really only thinking on the first one; after that, you're basically copying what you did on the previous problem. But if you do ten problems today, and then ten again tomorrow, and then ten later in the day, you will probably remember the steps much more clearly.

Math is a skill, and it's best learned the way you learn other skills, like playing a sport or a musical instrument. Just like you can't abstain from exercise all week and then complete eight hours of football practice on Sunday without an issue, you really shouldn't try to cram for math.

FACT

Treat your homework like what it is: preparation for the test. Homework is designed to give you a chance to practice what you learned, so the more you treat it like a test, the better.

Try It with Numbers

The biggest difference between pre-algebra and the math you've done before pre-algebra is the presence of variables (letter symbols that represent quantities in math). Sometimes it's hard to learn too many new things at once, and the variables can be just an added burden. So, try replacing the variables in a problem with numbers, and see if you can make sense of it that way!

When it comes to understanding new concepts, practice them with numbers first, and see what happens. Look at the solution and see what the variable equals, and then plug that answer in for the variable. Then, try manipulating the equation or expression so that you can understand how the numbers relate to one another. It might help you understand the rules.

Lots of concepts can be tested with numbers to see what the rule must be. For example, when you can't remember an exponent rule, try out a number and see what the rule must be. It's easier for our brains to work with numbers than with variables, because you encounter numbers in the real world all the time. Sometimes it will be easier to learn a new concept or test a rule with numbers, and then you can re-introduce variables once you've got the concept down.

Write the Test Yourself

Think about the test from your teacher's point of view. What would he or she want to include? What has been emphasized in class? (This is one reason it's really important to take good notes!) Probably the ideas and concepts that have been emphasized in class will be tested on the exam.

Take a look at the main ideas and concepts in the chapter being tested. What are the chapter headings? What are the terms you're expected to know? Between class and the text, you should have a pretty good idea of which main concepts are going to be tested.

Now, figure out how you'd test those concepts if you were writing the test. What kinds of questions would you include? What sort of tricks and traps would you put on there? What would you want to make sure your students know?

Once you've done all that, you basically have the test. What a great advantage: you have the test the day before the test! You'd be surprised how much of the time you get it right. Then you can study for the test you've prepared and make sure you know how to do every question on there.

Not only will this help you to study, but it will also make it easier for you to recognize what the questions on the test are asking you to do. It's a huge leg-up when you have taken the time to consider what's going to be on the exam, and you'll be surprised how well it prepares you.

Memorize or Learn?

In short, you should do both. There are some teachers who heavily emphasize memorization and really push you to use flashcards or similar tools to help you commit a list of facts and rules to memory. Other teachers encourage students to try to learn everything on a conceptual level.

Both teachers are right. You have to do both. You have to memorize the rules and the basics so that you can access them whenever you want. But you can't memorize skills—you have to learn them. When you memorize things, you will have them at your disposal to work with at any time. You're basically creating a toolkit for yourself. But the purpose of tools is to use them, which you will do by practicing your math. When you practice applying the concepts and rules you're studying, you learn them. Once you learn them, they stick with you.

Memorization works best in short stints. If you are trying to memorize something, try looking at it each morning, or a few times throughout the day for just a minute. Taping information to your bathroom mirror or the front of your notebook can help it stick in your head.

Memorization is necessary. You're not still sounding out each word you read as you move through this book; you've memorized different words, but it is important to note that you memorized them by learning them. You didn't put every word you know on a flashcard and memorize it. You started by memorizing a few short words and sounds; you read a lot; and this activity cemented some words in your brain.

Make flashcards for the rules and facts you want to know, but also practice applying them. For example, start by memorizing the rule for multiplying numbers with exponents. Even if you aren't sure why it is true, memorize it. Then practice it and use it in solving problems, and you will start to get comfortable with it. When you get confused, try it with numbers instead of variables to prove that you are remembering the rule correctly. Eventually you won't rely on the memorized rule anymore. Instead, it will come naturally because you've practiced it. That's real learning.

Some Things to Memorize

Memorization is a helpful first step toward learning. You can make flashcards or notes and keep an ongoing stack of things you want to memorize. If you keep them accessible and spend a little time each week reviewing the things you want to remember, they'll stay with you.

Every pre-algebra class is a little bit different. Different teachers, different textbooks, and different state standards mean that there's a slight variation in what's covered in pre-algebra classes all over the country and the world. But in general, here are some things that are good candidates for memorization:

- **Multiplication tables, one through twelve.** Don't rely on a calculator—when you memorize multiplication tables, it's much easier to find common factors in numbers.

- **Definition of terms (be on the lookout for bold terms in your textbook).** Terms like *integer, rational number, ratio, composite number,* and *polynomial* are terms you have to know.
- **New symbols.** Symbols like parentheses, root symbols, inequality symbols, and pi are things you want to recognize on sight.
- **Properties of fractions and exponents.** Memorization isn't enough here, but it's a good idea to memorize the rules for adding, subtracting, multiplying, or dividing fractions, as well as the rules for manipulating exponents.
- **Geometry definitions and formulas.** If your pre-algebra class includes geometry, memorizing will help you quite a bit. The definition of shapes such as triangles and cubes, and formulas such as those for the area of a rectangle or perimeter of a circle, will be worth committing to memory.
- **Properties of math.** The order of operations and the communicative, associative, distributive, and identity properties are all properties you need to know how to apply. Memorize them by name and then practice them until you are comfortable using them in various problems.
- **Names of place values.** Mixing up the *tens* digit and the *tenths* digit can get you in big trouble and result in the wrong answer.
- **Anything you keep messing up.** Even if it seems like a silly point or a specific detail that can't be memorized, if it's something that keeps tripping you up, put it on a flashcard and practice it every day until it's solid.

CHAPTER 2

Types of Numbers

Before you learn the details about what to do with different numbers, it's important to make sure that you are clear on some of the ways numbers can be classified. These clarifications are valuable and important when it comes to learning math. Most likely, you already understand many of the concepts discussed in this chapter without even realizing it. After all, you've been using numbers every day for the majority of your life. The explanations in the following sections just put a name to these concepts so you can easily refer to them as you continue your pre-algebra learning.

Counting Numbers

Up until now, most of the numbers you've dealt with have been what are called the **counting numbers** or **natural numbers**. These words mean the same thing. Counting numbers are the numbers you use to count: 1, 2, 3, 4, 5, etc. But as you know, there are a lot of other types of numbers, and they react in different ways when you do math with them.

Positive Integers

Another name for counting numbers or natural numbers is **positive integers**. (If you think it's confusing for them to have so many different names, many people would agree with you. Unfortunately, they do have different names, but it's unlikely you'll have to know any of these names except *positive integer*.)

Positive integers are the numbers you can use to count how much of something exists. For example, if you want to know how many people are in a room, you can count them and get a number. You probably already know that the number you come up with won't be four-and-a-half people or 6.8 people—it will be a whole number, such as five or nine.

Integers

What are integers? In short, integers are numbers that are not fractions or decimals. All counting numbers are integers. Counting numbers don't have any decimals on the end. They aren't fractions. They are whole numbers, without any extra pieces and without any pieces missing. Luckily, they are also called **whole numbers** for this very reason. Think about counting people in a room. You can only have whole people, not half people or quarters of people. That's why you count them with whole numbers.

Beyond Positive Integers

However, integers also include numbers that you don't encounter much in the real world. They include *all* the numbers that aren't decimals or fractions, not just the positive ones. That means when you hear someone talk about integers, they are including zero as well as all the negative integers.

Don't forget that zero is an integer! Zero isn't positive or negative, but it *is* an integer. It's also an even number, because it can be evenly divided into two groups that both have zero in them.

You can't really count negative integers in real life. After all, you can't look into a room and count –2 people in there. But negative numbers are important even though they aren't something you can see with your own eyes. Many numbers you use exist to keep track of concepts; they relate to real-world values even though you can't see them.

Think about money for a moment. Pretend you have $20 in your bank account. You aren't actually looking at the $20 when you count it. You're looking at the number on a page and knowing that it gives you information about how much money the bank is holding for you.

Say you make a mistake and swipe your debit card for a $30 purchase. Your bank account will now show a balance of –$10. Negative $10 isn't really something that can exist, but you know what it means. It means you have something even lower than zero. You have debt. You owe the bank $10. That is shown with negative numbers.

You'll be doing a lot of math with integers in pre-algebra. It's important for you to remember that zero and negative numbers are included in this list, because then you can make sure you understand all the rules that go with them.

Rational Numbers

There is a group of numbers called **rational numbers**, which gets its name from the word *ratio*. Rational numbers can be written as fractions. Some rational numbers are integers. In fact, all integers are rational numbers, because if you wanted to, you could write them as fractions. For example, think about the integer 5. If you wanted, you could write it as the fraction $\frac{5}{1}$. Or, you could write it as $\frac{10}{2}$. You can write any integer as a fraction, so all integers are rational numbers.

Before pre-algebra, the majority of the math you've done in life so far has been done with integers. But integers aren't the only rational numbers! Think about the integers 1 and 2. If you were counting people, you could have one person, or two people, but nothing in between. But that's because people can't be divided into pieces. However, lots of things *can* be divided into pieces. Think about money again, for instance. You already know there are a lot of values between $1 and $2. Maybe you express them as decimals and say that you have 1.5 dollars, or $1.50. Maybe you express them as fractions and say you have one-and-a-half dollars.

Why think about money?
Money is one of the most important ways that you actually use math in your daily life. You have to use ratios, fractions, decimals, negative numbers, and equations a *lot* in the real world to deal with money.

Even though all those values between 1 and 2 aren't integers, they are still numbers, so you can't forget about them. And there aren't just 100 values between 1 and 2. There is an *infinite* set of values between 1 and 2. There's 1.5 and 1.51 and 1.511 and 1.5111 and . . . you get the idea. Any of those values can be written as a fraction and is therefore part of the set of numbers called rational numbers.

Positive, Negative, and Zero

When you talk about positive numbers, you are talking about any number bigger than zero. It could be an integer, such as 1 or 1,000,000,000. It could be a non-integer, such as .5 or 100.5. If it's bigger than zero, it's positive.

When you first learn subtraction, a word problem might say, "Sally has five oranges, and she sells two of them. How many oranges does she have left?" Your teacher shows you five oranges on the table; she gives away two; and you can count that there are three remaining. You can see it with your own eyes and use counting numbers to prove that it's true. So it makes sense.

But what if the word problem said, "Sally has three oranges, and she sells five of them. How many does she have left?" Well, it wouldn't make much sense, because Sally can't sell more oranges than she has. Except for the fact that this is something stores do every day.

People buy things with money they don't have yet, and stores sell things that they don't yet have in stock. Everyone uses negative numbers to keep track of things that are missing. If you think of money as a stack of dollar bills, the least amount you could have would be zero. But when you think of it on paper or on a computer screen, you know that you can have less than zero, because it is possible that you owe money to someone.

Zero isn't positive or negative. That's because, as a number, it isn't bigger than zero and it isn't smaller than zero. It's nothing. It takes up no space. It's not something you have, and it's not something you owe. It just isn't there.

Like negative numbers, zero isn't really something you can see with your own eyes or hold in your own hands, which is why it can be hard to learn all the rules about this type of number. But you really need the concept of zero to have a math system that works to express all the things you want to express. You don't always have a tangible number of whatever it is you are trying to learn about—sometimes you have none, so you need zero to express this lack of quantity.

The Number Line

The number line is a visual tool that is used to show how big numbers are in relationship to one another. It's probably easiest to think about the positive side of the number line first, since you can visualize the numbers you see there easily.

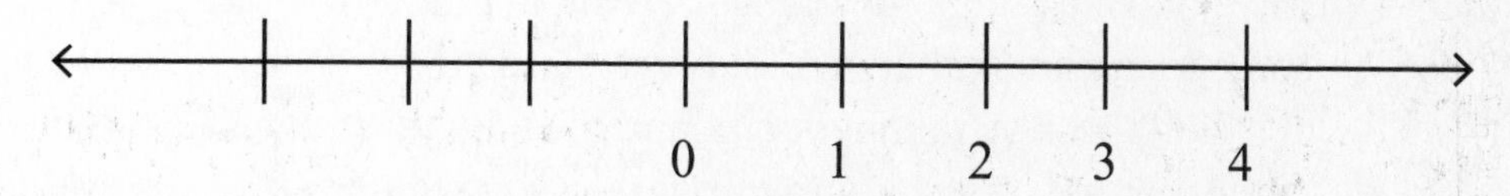

The number line works left to right, just like reading. Numbers on the right are always bigger than numbers on the left. The farther right you go, the bigger the numbers get. You already know that five is the biggest of those numbers, so it probably makes visual sense to you.

The number line goes on forever in both directions. As you move farther right, the numbers get bigger and bigger and bigger. The number line doesn't just include positive numbers. You can tell by those arrows drawn on the end of the number line that it is truly a line: it goes to infinity in both directions.

Now take a look at a bigger chunk of the number line.

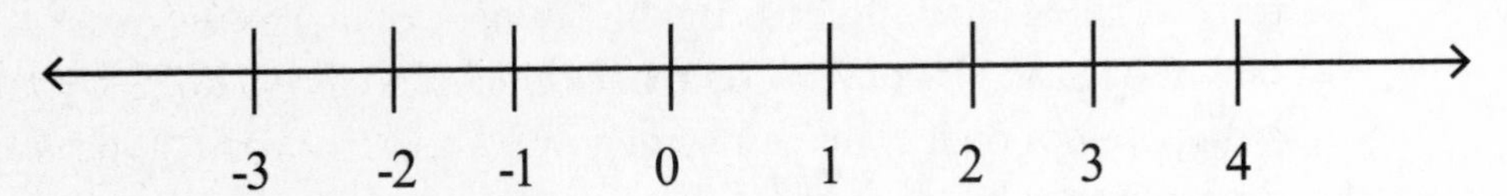

Now you can see that the number line keeps going to the left, out of the positive numbers, through zero, and down into the negative numbers. By using the number line, you can see how far apart or close numbers are, and you can also see which numbers are the biggest (the farthest to the right) and which ones are the smallest (the farthest to the left).

Size Versus Magnitude

Thinking about the **size** of negative numbers is kind of tricky. It might be best to start by thinking about the size of positive numbers. That's something you're probably already good at, because you've been doing it your whole life.

Bigger numbers, or numbers whose size is larger, represent more of something than smaller numbers do. For example, 10 is larger than 5, because ten oranges would be more than five oranges. Sometimes, a number can look bigger, but it's actually smaller. For example, the number 5.6789 takes up more space on the page than the number 6, but when you think about it in terms of quantity, six oranges are more than 5.6789 oranges. Thus 5.6789 oranges are more than five, but less than six.

With positive numbers, size and magnitude are the same thing. **Magnitude** tells you how far away from zero something is. For positive numbers, that is the same as saying how big they are. For example, the number 5 has a size of 5 and also a magnitude of 5. If you were to make that number smaller, the magnitude would also get smaller, as the number got closer to zero.

Because magnitude is a measure of distance, it will always be positive. Think about it this way: Bill walks one mile to his friend's house. Then he turns around and walks home. Bill's walk home is still one mile, even though he's going the other way. It doesn't become negative one mile just because he's changing directions. So measures of distance are always positive, and magnitude is a measure of distance. It just tells you how far away a number is from zero. (We also call magnitude **absolute value**.)

With negative numbers, things are different. You still determine size based on where things are on the number line: the farther right you go, the bigger numbers get. This means that –10 is actually *smaller* than –5, because it's farther left on the number line.

To conceptualize this idea, it can help to think of the number line as your bank account balance. You know that $100 is a bigger balance than $50. But now think about the negative side: would you rather have –$100 or –$50? (In other words, would you rather owe someone $100, or owe $50?) You'd rather have –$50, because owing someone $50 isn't as bad as owing them $100. In other words, –$50 is more than –$100. And that's because –50 is bigger than –100.

When it comes to negatives, the closer you get to zero, the bigger the number gets. That makes sense when you think about it. Closer to zero is closer to positive, and positive numbers are bigger than negative numbers. Think about extreme values, such as negative one billion versus negative one. Owing $1 is much better than owing $1 billion! So in this way, negative numbers are sort of "backwards" from positive numbers. When a negative number "looks" smaller, the number is actually bigger.

That's one reason we have the concept of magnitude, or absolute value, and why it's so important. Magnitude tells you how far a number is from zero. Therefore, negative 1 billion has a magnitude of 1 billion, because it's 1 billion away from zero. Negative one, on the other hand, has a magnitude of one, because it's only one away from zero. So even though negative 1 billion is a *lot* smaller than negative one, its magnitude is a *lot* bigger.

As negative numbers get smaller, their magnitude gets bigger. Their size gets smaller because they are farther to the left on the number line, and their magnitude gets bigger because they are moving farther away from zero on the number line.

Rational and Irrational Numbers (Real Numbers)

There are a couple more types of numbers to talk about before you really get into things. So far, you've learned about rational numbers. Rational numbers (all numbers that can be written as a fraction) include all the integers, plus all the non-integers that can be written as a fraction. That means 1, 2, 3, 4, and 5 are rational numbers. So are 1.5, 2.3, and 3.888. So are $\frac{1}{4}$, $\frac{1}{9}$, and $\frac{3}{16}$.

You can put any rational number on the number line. It has a real value here in the world, and it can be placed in order. A rational number is either negative, positive, or zero. It goes to the right of numbers it's bigger than, and to the left of numbers it's smaller than. You can find its spot in the order and place it on the number line.

But there are some numbers you can put on the number line that are *not* rational numbers. These are **irrational numbers**. Together, rational and irrational numbers make up the set of numbers called **real numbers**. A real number is any number you can put on the number line. Fortunately, real numbers are the last new category of numbers that you will learn about in pre-algebra.

Simply put, a real number is anything you can put on the number line. If you can write it as a fraction, it's a rational number. Most of the numbers you'll deal with in this book are rational numbers. Irrational numbers are numbers that can't be written as a simple fraction. One of the most famous examples of an irrational number is pi, which is represented with a Greek letter that looks like this: π.

Pi is a number that comes from the mathematical relationships found in circles, and it is needed to answer questions about circles. But it has to be written as a symbol or as a mathematical estimate, because it can't be written out as a fraction since, in decimal form, it goes on forever with no sort of pattern.

Square Roots

Most of the other irrational numbers you will find in this book will be the square roots of integers (a square root is a number that produces a specified quantity when multiplied by itself; for example, $2 \times 2 = 4$, so 2 is the square root of 4). Some square roots, like the example just given, are integers. However, other square roots are not. For example, if you take the square root of 2, you'll get a number that can't be expressed as any sort of fraction. As a decimal, it just goes on forever and ever without any sort of pattern. That makes it an irrational number, and means you would usually just keep writing $\sqrt{2}$ instead of trying to simplify it further, because you can't do that without estimating.

A number whose square root is an integer is called a **perfect square.** So, for example, 100 is a perfect square because its square root is 10. The first ten perfect positive perfect squares are 1, 4, 9, 16, 25, 36, 49, 64, 81, and 100.

If all this has your head spinning, don't fret! Most of the time, you don't need to remember the definitions of types of numbers. But it's good for you to know them from the start so that you can talk about numbers more clearly. The following chart sums up the types of numbers that have been mentioned thus far.

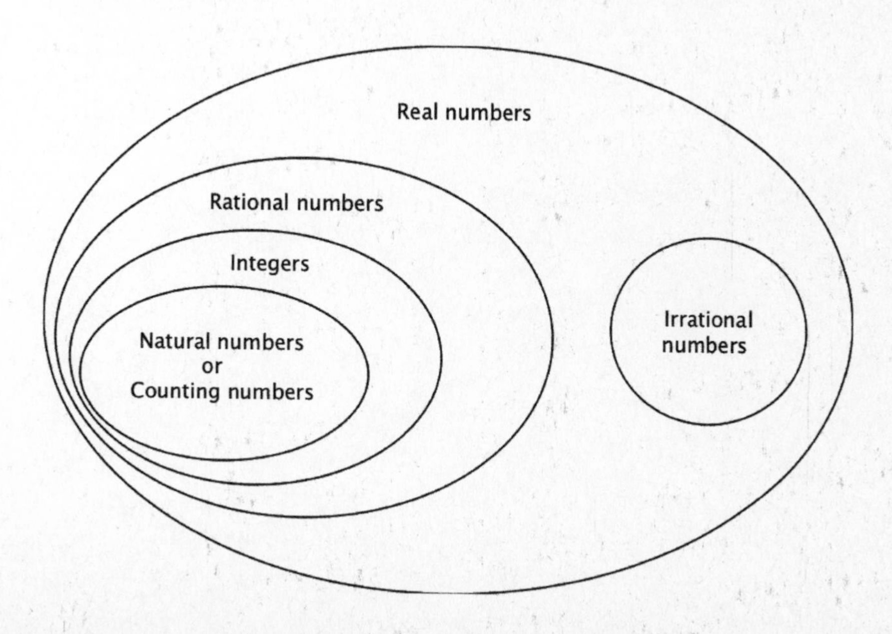

Chapter 2 Exercises

Circle the integers: 0 .5 2 –4 ½ 100 6.7 –70 12

Every number except zero is positive or negative, and every number (including zero) is either an integer or a non-integer. Indicate whether each of the following numbers is a positive integer, a negative integer, a positive non-integer, or a negative non-integer.

1. –2
2. 5
3. –.5
4. 27
5. .38
6. –30
7. 1
8. 1.5
9. $-\frac{1}{2}$

CHAPTER 3

Factoring

This chapter is all about factoring: what factors are, how they work, how they can be written, and what they can be used for. Factoring skills are important for a number of reasons, but they are particularly important in working with fractions and when simplifying algebraic expressions and equations. In other words, being good at factoring can make hard things a whole lot easier.

Multiplication Tables

Factoring is an important skill in mathematics. It's been important since you started learning math in kindergarten, and it will be important for as long as you keep studying math.

Multiplication tables are crucial for factoring properly. If you don't know your multiplication tables at least up through the tens table, take the time to learn them. If you are struggling with math, it could be because you aren't recognizing patterns or shared factors since you aren't comfortable enough with your multiplication tables. Multiplication tables *will* help you, even if you are allowed to use a calculator on exams and even if they won't be explicitly tested in class. They are worth learning.

	1	**2**	**3**	**4**	**5**	**6**	**7**	**8**	**9**	**10**	**11**	**12**
1	1	2	3	4	5	6	7	8	9	10	11	12
2	2	4	6	8	10	12	14	16	18	20	22	24
3	3	6	9	12	15	18	21	24	27	30	33	36
4	4	8	12	16	20	24	28	32	36	40	44	48
5	5	10	15	20	25	30	35	40	45	50	55	60
6	6	12	18	24	30	36	42	48	54	60	66	72
7	7	14	21	28	35	42	49	56	63	70	77	84
8	8	16	24	32	40	48	56	64	72	80	88	96
9	9	18	27	36	45	54	63	72	81	90	99	108
10	10	20	30	40	50	60	70	80	90	100	110	120
11	11	22	33	44	55	66	77	88	99	110	121	132
12	12	24	36	48	60	72	84	96	108	120	132	144

Factors

Only integers have factors, and all integers have factors. **Factors are the integers you can multiply together to get another integer.** For example, the factors of 12 are any integers that you can multiply by another integer to get 12. You will start with the easiest pair of factors. Every integer greater than 1 has at least 1 and itself as factors. So in terms of the number 12, for example, one pair of factors is 1 and 12. Because of this fact, every integer except for the number 1 has at least two distinct factors: 1 and the number itself. (For the number 1, those are the same, so it only has one distinct factor.)

However, you know that 1 and 12 are not all the factors of 12. What other integers can you multiply together to get 12? You can multiply 2 and 6 to get 12, so 2 and 6 are also factors of 12. You can also multiply 3 and 4 to get 12, so 3 and 4 are also factors. So now you have six different factors for 12: 1, 2, 3, 4, 6, and 12.

Take note of the characteristics of these factors. First, they are all equal to or smaller than 12. When you find the factors of an integer, you only have to consider the integer and those numbers smaller than it. Notice something else: they work in pairs. A factor of an integer is a whole number that you can multiply by another whole number to get that integer. That means that each factor has to have a sort of partner that it multiplies with. So you could list the factors of 12 like this:

1 and 12
2 and 6
3 and 4

Each of these pairs multiplies to 12. The factors of an integer will always pair up like this. If you have to list all the factors of a number, making a chart of factor pairs can be the fastest way to do so.

Finding Factors

When you list out the pairs like this, you can be sure that you've included every possible positive factor. How do you do it? Use the number 100 as an example, and walk through the steps.

1. Start with 1 and the number itself. That's always the first pair of factors. So the first pair of factors in your factor pairs chart would be 1 and 100.
2. Moving up from the number 1, check to see if each of the next integers (2, 3, 4, etc.) is a factor. If it is, find its factor pair and write the pair down.

So you know that 1 is a factor. What about 2? Yes, 2 is a factor of 100, because $2 \times 50 = 100$. If you didn't know this was true, you could check. You could divide 100 by 2 and see whether you get an integer as an answer. If you did, that integer and 2 would be factors. If you didn't, then 2 wouldn't be a factor.

Because 2 is a factor, and $2 \times 50 = 100$, 50 is also a factor. So you've found your next factor pair! Now your factor pair list looks like this:

1 and 100
2 and 50

3. Keep trying bigger integers to see if they are factors. If they are not, skip them and keep moving. If they are, find their factor pair and write both factors down.

So in this example, the next number to try is 3. Is 3 a factor of 100? No, it's not. You might realize this because 3 is a factor of 99, so it can't be a factor of 100. Or you might just know that you can't take a dollar (100 cents) and divide it evenly among three people. Since it doesn't divide evenly, 3 isn't a factor.

That means you can skip 3 and go on to 4. Four *is* a factor of 100. It gives you 25 as a factor as well, because $4 \times 25 = 100$. If you didn't know that, you could find 25 by dividing 100 by 4. So now your factor pair list looks like this:

1 and 100
2 and 50
4 and 25

Do you notice what's happening? The factors on the left side of the list are getting bigger, while the factors on the right side of the list are getting smaller.

4. So on you go, testing more numbers. What about 5? Yes, $5 \times 20 = 100$. What about 6? 7? 8? 9? None of those are factors of 100. What about 10? Yes. $10 \times 10 = 100$, so 10 is a factor of 100.

Now you have a complete list, which looks like this:

1 and 100
2 and 50
4 and 25

5 and 20
10 and 10

You'll know you are done when you don't have any numbers left to try. In this example, all the factors bigger than 10 are already listed on the right side of the list, so there can't be any other factors. And now you know that the complete list of factors of 100 is 1, 2, 4, 5, 10, 20, 25, 50, and 100.

Try going back to the number 12 as another example, and work through the steps to make sure you've got it.

First, list 1 and the number itself: 1 and 12. Then, continuing on up from 1, try each integer to see which ones are factors of 12. You know that 2 is a factor, and that tells you that 6 is also a factor (since $2\times6=12$). You know that 3 is a factor, which tells you that 4 is a factor (since $3\times4=12$). Then you can stop because the next number you would try is 4, and that's already listed in the right-hand column of our factor pairs list. So now you can rest assured that you have a complete list.

Divisibility

An integer is **divisible** by its factors. That means that 12 is divisible by 1, 2, 3, 4, 6, and 12. The word *divisible* refers to integers, and means that an integer can be divided evenly into a certain number of groups. You can divide twelve people into one group, two groups, three groups, four groups, six groups, or twelve groups. You can't divide them into five or seven or ten groups without things getting really, really messy.

An integer is divisible by its factors; it's not divisible by anything else. That's what factors are—numbers that an integer is divisible by. You might notice that this chapter only talks about integers. That's because integers are the only kind of numbers that have factors, or that are divisible evenly by other integers.

You already know how you can test to see if a number is divisible by another number. For example, when you tried to figure out if 100 was divisible by 3, you tested it by dividing 100 by 3 and realized that the result was not an integer. You can always do that. If you divide a number by another number and don't get an integer, that first number is *not* divisible by the second number. For example, when you divide 50 by 5, you get 10. So 50 is divisible by 5 (and 10). But when you divide 50 by 4, you get 12.5 (or 12 with a remainder of 2). Since 12.5 isn't an integer, you know that 50 is *not* divisible by 4.

If you know your multiplication tables, you'll be able to glance at any two-digit number and find its factors pretty quickly by working through your multiplication tables. This process helps with big numbers, too. Once you divide them by 2 or 3 or 10, they'll become smaller, and then your multiplication tables will come right back into play.

Divisibility Shortcuts

There are some shortcuts that might help you glance at a number and decide if it is divisible by another number. Here are a few shortcuts that are particularly useful for finding factors when making a factor pair list:

Divisible by 1: Every single integer is divisible by 1. See? That one was really easy.

Divisible by 2: Every even integer is divisible by 2, and every odd integer is not. (In fact, *even* actually means "divisible by 2.") At a quick glance, you can look at the ones digit, or unit's digit, of a number. If the ones digit is 0, 2, 4, 6, or 8, that's an even number, so you know that 2 will be a factor.

Divisible by 3: This is a weird trick, but it's nice to know. If you want to tell if a number is divisible by 3, just take all the digits and add them up. If their sum is divisible by 3, the number is divisible by 3. For example, if you aren't sure whether 36 is divisible by 3, just add 3 and 6. They add up to 9; since 9 is divisible by 3, 36 is divisible by 3.

Divisible by 4: You know that 4 is just 2×2. This means anything that is divisible by 2 twice will also be divisible by 4. So when you divide the number by 2, see what you're left with as the partner factor of 2. If that number is even, then the original number must be divisible by 4. For example, check out 100. One hundred is even because it ends in 0. When you divide it by 2, you get 50. Since 50 is also even (because it ends in 0), you know that 100 must be divisible by 4.

Divisible by 5: This is an easy one. Take a look at the ones digit of the number. If it's 5 or 0, the number is divisible by 5. If the ones digit is anything else, the number is not divisible by 5. (This should make sense to you if you think about your fives table; all the numbers in that table end in 0 or 5.)

Divisible by 6: If you've got 2 and 3 in your factor list, you've got to have 6 as well. If a number is divisible by 2 and by 3, it's also divisible by 6, since 6 is just 2×3.

Divisible by 7: Unfortunately, there is no nifty trick for 7. If you aren't sure whether a number is divisible by 7, you just have to do long division and check.

Divisible by 8: Since you know that $8 = 4 \times 2$, you can forget about a number being divisible by 8 if it isn't divisible by 4. If it *is* divisible by 4, check out the factor pair for 4. If that other factor is even, then the number must be divisible by 8, because the number has 4 and at least one other 2 as a factor. Take 72, for example. You can tell at a glance that it's divisible by 2, because it ends in 2. When you divide 72 by 2, you get 36, which is even. Since you can make 72 by multiplying 2 by an even number, you know it's divisible by 4. So you divide 72 by 4, and you find out that 72 divided by 4 is 18 (so $4 \times 18 = 72$, which means 4 and 18 are both factors of 72). Since 72 can be written as 4×18, or 4 multiplied by an even number, it means 72 must be divisible by 8. And it is: $8 \times 9 = 72$.

Divisible by 9: The trick for checking for divisibility by 9 is pretty much the same as the trick for 3. Add up all the digits of the number, and if their sum is divisible by 9, the number is divisible by 9. For example, take a look at 81. Because $8 + 1 = 9$, and 9 is divisible by 9, you know that 81 is divisible by 9.

Divisible by 10: If a number has zero as the ones digit (meaning it ends in zero), that number is divisible by 10. If it doesn't end in zero, it's not divisible by 10.

Multiples

The easiest way to think about multiples is to think about the first way you learned them: again, the multiplication tables. Think about your threes table, for example. It gives you 3×1, 3×2, 3×3, and on and on. In other words, it lists all the **multiples** of 3.

A multiple of 3 is 3 times any integer. **A multiple of an integer is that number times any other integer.** For example, 3, 6, 9, 12, 30, and 3,000 are all multiples of 3. There are infinite multiples of a number. You can't list all the multiples of 3 in the same way you can list all the factors of 3. When you list the factors, it's a finite list. When you list the factors of 100, all the numbers on that list are equal to or smaller than 100, and at some point, you will run out of numbers to list. But if you wanted to list all the multiples of 100, you'd have to keep writing for your whole life, and you still wouldn't list them all. You'd write down 100, 200, 300, 400 . . . and then just keep going forever and ever!

Everything Is Related

The concepts of divisibility, factors, and multiples are all related. You can probably already see that—you can't really talk about factors without talking about multiplication, and you can't really talk about divisibility without talking about factors.

This relationship means there are lots of ways to say the same thing. For example, you can say:

Twelve is a multiple of four.
Twelve is divisible by four.
Four is a factor of twelve.

These are just different ways of saying the same thing. The first sentence tells you that four times a number will give you twelve. The second sentence tells you that twelve can be broken into four even groups of some amount. And the third one tells you that twelve can be written as four times some number. Do these rules sound sort of similar to you? They should, because they tell you exactly the same thing. This actually makes it easier to ask clear questions and give clear answers.

Zero is a multiple of every integer, because you can multiply any integer by zero to get zero as the product. Zero is not a factor of any integer, because nothing is divisible by zero.

Simplifying Using Common Factors

You will often use common factors to simplify fractions. Actually, you'll use the existence of common factors to do a *lot* of things to fractions. You need to think about common factors when you simplify, add, subtract, multiply, or divide fractions.

For now, focus on how you can simplify fractions using the existence of common factors. When a fraction is written in its simplest form, the numerator (the top number) and the denominator (the bottom number) will not have any prime factors in common. In other words, they won't have any common factors other than the number 1.

For example, the fraction $\frac{1}{3}$ is written in its simplest form, because 1 and 3 do not have any common factors other than the number 1. You can say the same thing about $\frac{2}{3}$ or $\frac{5}{7}$. Both of these fractions are in their most simplified forms: 2 and 3 don't share any prime factors, and neither do 5 and 7.

When the numerator and denominator *do* share common prime factors, you can simplify the fraction by eliminating the factors they have in common. Consider the fraction $\frac{2}{4}$. Two quarters is the same as one half. But how do you get there?

You can remove any factor that the numerator and the denominator have in common. In the fraction $\frac{2}{4}$, the numerator and the denominator both have a 2 in them. In other words, think of this fraction as $\frac{(1\times2)}{(2\times2)}$. You can cancel out the 2 that is shared by both the numerator and the denominator, which leaves $\frac{1}{2}$.

Once a factor is in the numerator and the denominator of a fraction, it stops having any effect on the size of the fraction. Take the 2 you just canceled out in the previous example. Multiplying the denominator by 2 gives you twice as many pieces of the pie. Multiplying the numerator by 2 lets you take twice as many pieces. So, it's exactly the same to take one out of every two pieces, or $\frac{1}{2}$ the pieces, as it is to take two out of every four pieces, or $\frac{2}{4}$ of the pieces.

That's the conceptual reason you can cancel out any common factors in the numerator and denominator of a fraction without having an effect on the size of the fraction. But what's the mathematical reason?

$\frac{(1\times2)}{(2\times2)}$ is the same as $\left(\frac{1}{2}\right)\left(\frac{2}{2}\right)$. You can see that the fraction $\frac{2}{2}$, with a 2 in the numerator and a 2 in the denominator, is equal to 1. Think about it: two pieces out of two pieces is one whole thing. And because $\frac{2}{2}$ is the same as 1, multiplying by it doesn't do anything to change the size of the fraction. It just changes the way it's written.

If you are able to look at a fraction and notice a shared factor in the numerator and the denominator, that's great! But if you can't, you can start factoring the numerator and the denominator to see if that helps you to find a common factor. For example:

1. **Simplify $\frac{4}{20}$.** First, notice that 4 and 20 have 4 as a common factor. Because of that, you are able to divide both the numerator and the denominator by 4. If you were to write it out, the math would look like $\frac{(4\div4)}{(20\div4)}$. Divide the numerator by 4 and you are left with 1. Divide the denominator by 4 and you are left with 5. You're allowed to do that because when you divide the top and bottom by 4, you're dividing by $\frac{4}{4}$, which is the same as dividing by 1. It doesn't change the value of the number at all. Thus, you've simplified the fraction to $\frac{1}{5}$.
2. **Simplify $\frac{17}{50}$.** You know that 17 is a prime number, which means its only factors are 1 and 17. You also know that 50 is not a multiple of 17. Thus, you know that 17 and 50 don't share any common factors, so this fraction cannot be simplified any further.
3. **Simplify $\frac{24}{144}$.** You don't know all the common factors, but you do know both numbers are even, which means that they're both divisible by 2. Go ahead and factor 2 out of the numerator and the denominator by dividing them both by 2. By doing this, you are simplifying $\frac{(24\div2)}{(144\div2)}$,

and you will be left with $\frac{12}{72}$. Since you are not sure what their greatest common factor is, but you know that 12 and 72 are both multiples of 6, factor a 6 out of the numerator and the denominator by dividing them both by 6, $\frac{(12 \div 6)}{(72 \div 6)}$. Once you've done this, you'll be left with $\frac{2}{12}$. Once again, these are both even, so you can divide them both by 2, leaving you with $\frac{1}{6}$. It turns out that you were able to factor 24 out of both the numerator and the denominator. If you'd noticed that right away, you could have done the whole simplification in one step! But you can start with any common factor you recognize, and as long as you keep going until there are no shared factors left, you'll get to the right answer.

Finding common factors of the numerator and denominator is essential to simplifying a fraction. It's also very helpful for making fractions easier to work with.

Chapter 3 Exercises

Answer the following questions with "yes" or "no."

1. Is 50 divisible by 1?
2. Is 33 divisible by 10?
3. Is 7 divisible by 14?
4. Is 81 divisible by 1?
5. Is 0 divisible by 25?
6. Is 25 divisible by 0?
7. Is 28 divisible by 8?

Answer the following questions with "yes" or "no."

1. Is 30 a multiple of 10?
2. Is 15 a multiple of 30?
3. Is 0 a multiple of 9?
4. Is 9 a multiple of 0?
5. Is 1 a multiple of 16?

Answer the following questions with "yes" or "no."

1. Is 12 a factor of 24?
2. Is 1 a factor of 14?
3. Is 7 a factor of 21?
4. Is 4 a factor of 6?
5. Is 12 a factor of 6?

List the positive factors of each number.

1. 1
2. 6
3. 21
4. 17
5. 144

CHAPTER 4

Prime Numbers

This chapter covers factorization in more detail. Being able to break a number down into prime factors is an essential skill that will help you with pre-algebra problem solving. Primes are the building blocks of numbers, so being able to break a number into primes will help you perform any number of mathematical operations.

What's a Prime Number?

A prime number is a number that has exactly two positive factors: 1 and itself. So the first prime number is 2. The only factors of 2 are 1 and 2. That's exactly two factors and two factors only, so the number 2 is prime.

The first prime number is 2. It is also the only *even* prime number. That's because every other even number has 2 as a factor—that's what makes it even. So the rest of the prime numbers, other than 2, are odd. That certainly doesn't mean that every odd number is prime. What it does mean is that every prime number is odd. For example, 15 is odd, but it isn't prime, because it has more than two factors. It has 1, 3, 5, and 15.

To test if a number is prime, you have to try to factor it. If the only factors it has are 1 and itself, it's prime. The first twenty-five prime numbers are 2, 3, 5, 7, 11, 13, 17, 19, 23, 29, 31, 37, 41, 43, 47, 53, 59, 61, 67, 71, 73, 79, 83, 89, and 97. This list isn't something you have to memorize, but take a look at it for a minute and check out the numbers. Notice that they don't have a pattern. For a while it's every odd number, but then you skip a few, and sometimes you skip a bunch, and sometimes two odd numbers in a row make the list. Prime numbers don't have a clear pattern, and they don't run out. There are *huge* prime numbers. Mathematicians just discovered a prime number that is *17 million digits long*. You'd need over 6,000 pages of a book like this one just to write down the number!

The list above has all the prime numbers between 1 and 100. You should be able to test any number between 1 and 100 that is *not* on the list and prove that it is *not* prime; in other words, any number between 1 and 100 that is *not* on that list should have more than two positive factors. Try it and see!

One is not a prime number. Sometimes people remember the definition of a prime number as "a number that is only divisible by one and itself." That definition is almost right, but it's misleading. The real definition of a prime number is that it has exactly two positive factors: 1 and itself. You can see how that definition does *not* include the number 1.

The Factor Tree

You already know one way to find all the factors of a number: you can list them all out in factor pairs. But sometimes that isn't the most efficient way, and sometimes it doesn't tell you everything you want to know in the clearest way.

It's great to be able to find all the factors of a number, but it's also important to be able to list the prime factors. Prime factors are the simplest building blocks of integers. Just like atoms make up all the matter in our world and can't be broken into smaller pieces, prime numbers make up all the integers. They can't be factored any further than themselves; they are the end of the line.

So how do you break a number into its prime factors? You use a factor tree. Although it sounds fancier, it's actually a lot easier than making a list of factor pairs.

There is a term for integers that aren't prime. They are called **composite integers.** A composite number is an integer that is divisible by at least one factor other than 1 and itself. For this reason, all non-prime numbers except for 1 are composite numbers. The number 1 is neither prime nor composite.

Let's say you want to find the prime factors of 100. Start by breaking 100 up in any way that you find easiest. If you know that $100 = 10 \times 10$, for example, start there. You are going to use a **factor tree** to branch out from 100, breaking the pieces into smaller and smaller chunks until the end of each branch is a prime number. Then you can stop, because primes can't be broken down any further.

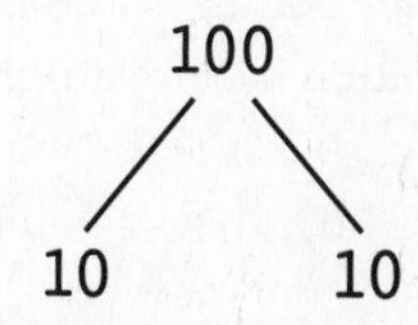

In this example, you are not finished because the "branches" of your tree are not yet prime numbers. So you can break them each down further.

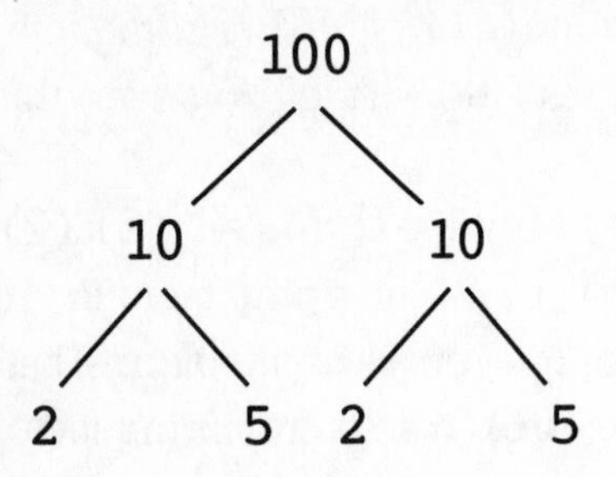

Now you can stop, because all of your branches end in prime numbers. Now you know that the different prime factors of 100 are 2 and 5, and that all the prime factors of 100 are 2, 2, 5, and 5.

If it helps, you can circle the primes once you reach them to help you recognize when you are at the end of a "branch." If you circled the primes, your prime factor tree of 100 would look something like this:

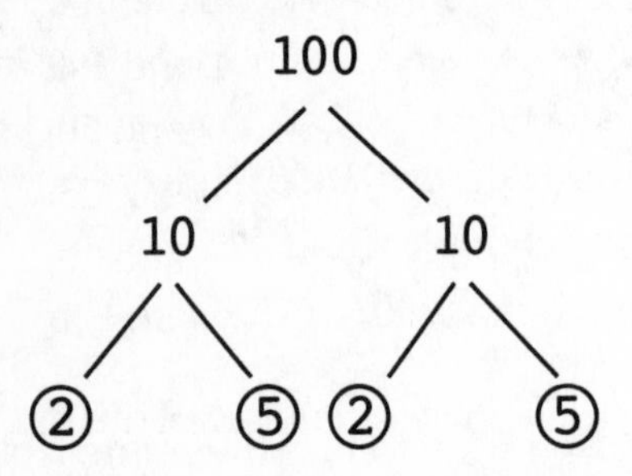

What's great about the factor tree is that it gives you a complete list of primes, and it doesn't matter where you start or how you factor the numbers. Imagine that you thought of 100 as 4×25 instead of 10×10. When the tree starts out, it will look pretty different:

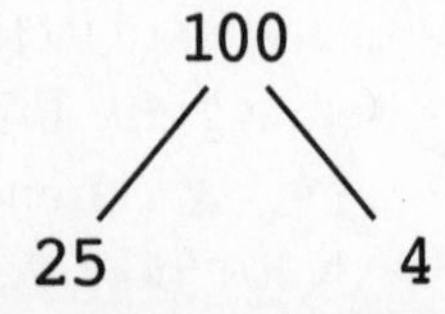

However, look what happens when you keep factoring until all of your "branches" end in primes:

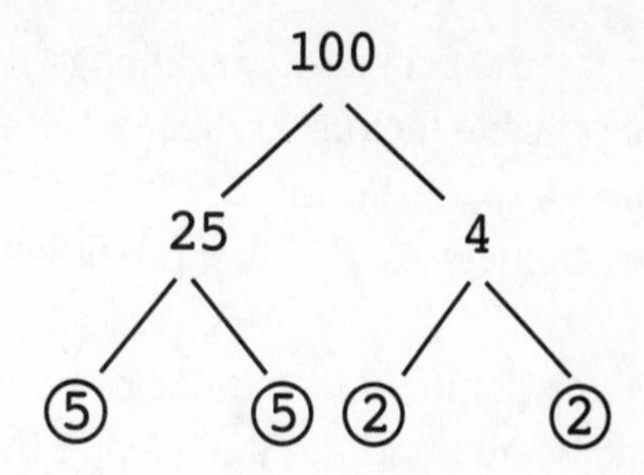

It ends in exactly the same place! That's the beauty of prime factors. You can only make one list of the prime factors of a number, because only one list of such prime factors exists.

Prime Factorization

Notice that you can make all the factors of 100 out of different combinations of 2, 2, 5, and 5. In fact, every pair of factors can be made by grouping the 2, 2, 5, and 5 in different clumps. Here is the old factor pairs list for 100:

1 and 100
2 and 50
4 and 25
5 and 20
10 and 10

If you wanted to, you could rewrite the list like this:

1 and $2\times2\times5\times5$
2 and $2\times5\times5$
2×2 and 5×5
5 and $2\times2\times5$
2×5 and 2×5

That's because every pair of factors for a number has to be made out of its prime factors and nothing else. Primes are the building blocks of integers.

An integer has exactly one prime factorization. As you saw while making factor trees, no matter how you start factoring an integer, you will always break it down into exactly the same primes. No other integer can share that same prime factorization. Every integer has its own unique prime factorization.

You can use the prime factorization of a number to help you quickly list out all of its possible factors. Try using 100 as an example again. You know that the prime factorization is $2 \times 2 \times 5 \times 5$. Because of that, you know that these prime numbers are the only possible building blocks of all the factors of 100.

So you can start making your factor pairs list, but this time you're armed with your factor tree and the knowledge that 2 and 2 and 5 and 5 are the only prime factors. For this example, making the list will go a lot faster!

As usual, start with 1 and the number itself. So your list starts off like this:

1 and 100

Now move on to 2, which you *know* is a factor because it's one of your favorite primes. Its partner factor will be the product of all the rest of the primes, $2 \times 5 \times 5$, which is 50. Alternatively, once you know 2 is on the list, you can just do $100 \div 2$ in your head and know that the factor pair for 2 is 50. Now you've got:

1 and 100
2 and 50

No point in thinking about the number 3. After all, 3 is prime, so if it were a factor, it would show up in your prime factorization. Since it isn't in your prime factorization, it must not be a factor.

What about 4? For 4 to be a factor of 100, you'd have to have two 2s in your prime factorization. You do! So 4 is a factor, and its partner is the rest of the primes multiplied together. Therefore, its partner factor is 25, since $5 \times 5 = 25$.

You also know that 5 is a factor because it's a prime, and it's on your list of prime factors of 100. You can then move in the same way through the rest of the list to end up with:

1 and 100
2 and 50
4 and 25
5 and 20
10 and 10

And now you can stop, because you know that all the factors bigger than 10 are already on your list.

See how prime factorization and the factor tree can help? While there may be a lot of ways to factor a number, there is only one way to show its prime factors. That's why you can look at the factor tree for 100 and be sure that 100 is not divisible by 8. Because the prime factorization of 8 is $2 \times 2 \times 2$, the *only* way a number can be divisible by 8 is if its prime factorization has three 2s in it.

Prime Notation

Now that you are able to use a factor tree to break a number down into its prime factorization, there's one other thing you might want to know: how to write an integer in prime notation. It involves exponents, but you don't have to be an expert on exponents to write an integer in prime notation.

Prime notation is just a helpful way to write out the prime factorization of a number. For numbers with only a few prime factors, it honestly doesn't help you that much. For numbers with a lot of prime factors, prime notation can make your prime factor information a lot easier to use and understand.

For example, take the number 100. You already know that the prime factors of 100 are 2, 2, 5, and 5. Therefore, you can write 100 as $2 \times 2 \times 5 \times 5$, but if you want to write 100 in prime notation, you group all the identical primes together. The goal of prime notation is to make the information easier to understand and use. Think about how you'd *say* the prime factors of 100. You probably wouldn't say, "There's a 2 and a 2 and a 5 and a 5." You'd probably find it easier to say, "There are two 2s and two 5s." That's exactly what prime notation does: it groups the identical primes together so that they are easier to talk about and use.

In this example, instead of writing 2×2, you would write 2^2. That's because the 2 is being multiplied by itself, which is what 2^2 stands for. In other words, the base, or bottom number, will be the prime factor you are counting. The exponent, or the little number, will be a count of how many times that factor shows up.

If you are at all familiar with exponents, you know that this is how exponents always work. They tell you how many times the bottom number, or base, will be multiplied by itself. In prime notation, the exponent tells you how many times the base is a factor in the multiplication.

The prime notation of 100, then, is $2^2 \times 5^2$. You can even write it without the $\times$ sign if you want, which would look like this: $2^2 5^2$. For now, leave the multiplication sign so you're only doing one new thing at a time.

Now consider the number 8. To write the prime notation, start by doing a factor tree.

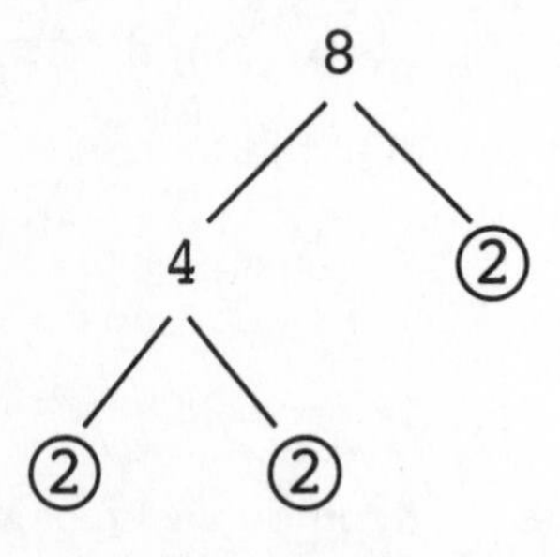

Now you know that $8 = 2 \times 2 \times 2$. There is only one distinct prime factor: 2. How many times does it show up? It shows up three times. Thus, you have the number 2 on the list three different times. So your prime notation of the factorization of 8 would be 2^3.

When using prime notation, there is no need to put the exponent of 1 on prime factors that only appear once. It isn't technically wrong, but mathematicians usually don't bother writing 1 as an exponent, because it's the same as having no exponent at all.

If you're asked to write the prime factorization of 120, for example, here are the steps you would follow.

1. Write out the factor tree and circle all the primes.

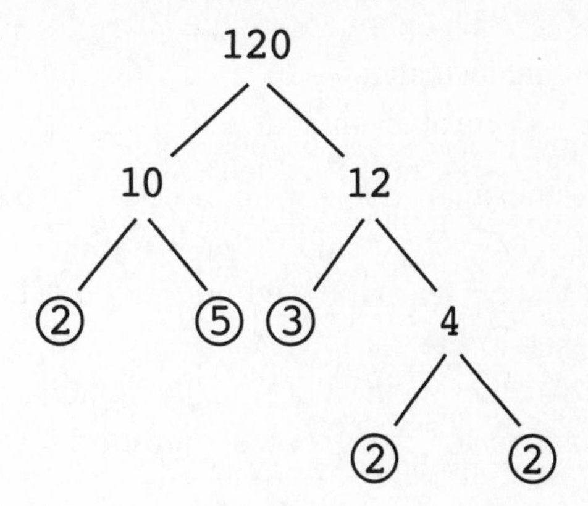

   Now you know that $120 = 2 \times 5 \times 3 \times 2 \times 2$.
2. Group the identical primes together. It is convenient to list them in order of their size, smallest to largest. Technically, this doesn't change the value of the number; it just makes the number easier to understand at a glance. $120 = 2 \times 2 \times 2 \times 3 \times 5$.
3. Write each distinct prime as a base and then raise each one to an exponent that expresses how many times that prime factor shows up. Thus, the prime factorization of 120 is $120 = 2^3 \times 3 \times 5$.

That's it! Like a lot of pre-algebra, learning to use prime notation doesn't require you to do any math you didn't already know. It simply requires you to remember the directions and learn a new way of writing something you already knew.

Remember: math is a code, and a good code requires that everyone using the code agree on what each symbol or collection of symbols means. So a lot of the confusion of math can be resolved when you get familiar with what the code tells you—and that comes with practice.

Greatest Common Factor

You could simplify $\frac{24}{144}$ by removing one common factor at a time, but this requires a lot of steps. Some people prefer to do the simplification step-by-step rather than doing it all at once. If you are one of those people, you can simplify your fractions that way. But even if you like that method better, you'll want to learn the method described here, because you may be explicitly asked to use this method on a quiz or exam.

This other method of simplifying fractions is called finding the greatest common factor. The **greatest common factor** (or **GCF**) is the largest factor shared by two numbers. It might be easiest to stick with the same example to see how it works. If you wanted to simplify $\frac{24}{144}$, you could start by factoring the numerator and the denominator to find all their common factors in one step. Check it out:

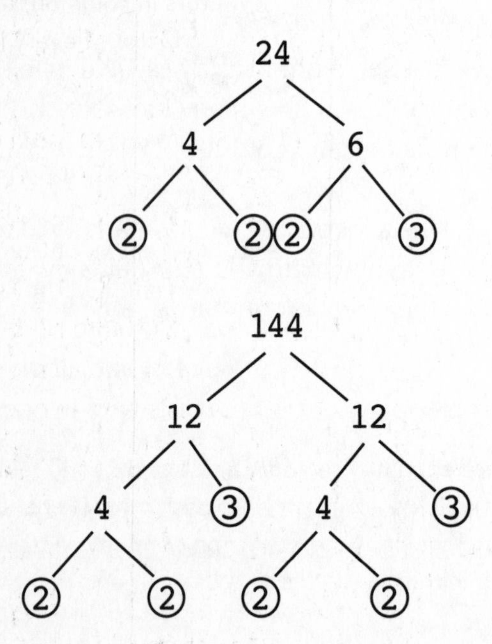

Now you know that you can rewrite $\frac{24}{144}$ as $\frac{(2\times2\times2\times3)}{(2\times2\times2\times2\times3\times3)}$. At a glance, both the numerator and the denominator share three 2s and one

3, so that's the complete list of their common factors. Any factor that shows up in the numerator and the denominator can be "factored out": in other words, you can divide both the numerator and denominator by that factor, and the factor will disappear (but the value of the fraction will stay the same, since you're effectively dividing by 1). Cancel the common factors out of the numerator and the denominator, and you are left with $\frac{1}{(2\times3)}$. Thus, you know that $\frac{24}{144}$ simplifies to $\frac{1}{6}$.

What you've done here is find the greatest common factor of 24 and 144. You found the biggest factor that they have in common—the largest number that factors into both 24 and 144.

The greatest common factor is sometimes called the **greatest common divisor**. Every two integers have a greatest common factor. If they don't have any prime factors in common, their greatest common factor is 1. That's because 1 is a factor of every integer.

You will need to find the GCF of two numbers when one of those numbers is the numerator of a fraction and the other number is the denominator. By finding the greatest common factor, you can simplify the fraction quickly.

It is possible to find the GCF by listing all the factors of two numbers and identifying the biggest factor that shows up on both lists. So you *could*, for example, write out all the factors of 24 and all the factors of 144 and look for the biggest number that your two lists have in common.

Factors of 24: 1, 2, 3, 4, 6, 8, 12, and 24
Factors of 144: 1, 2, 3, 4, 6, 8, 9, 12, 16, 18, 24, 36, 48, 72, and 144
Biggest number on both lists? 24. Thus, 24 is their GCF.

While this works, it is faster, easier, and better to use prime factorization to find the GCF. Find the prime factorization of each number. Then, look for the prime factors that the two prime factorizations have in common. When you multiply those prime factors together, you will get the GCF.

Prime factors of 24: $2\times2\times2\times3$

Prime factors of 144: $2\times2\times2\times2\times3\times3$

Prime factors they have in common? $2\times2\times2\times3$. Since that equals 24, 24 is the GCF.

This is one of the many reasons that primes are so great: they're useful. When you look at those full lists of factors, there is a lot of overlap. But with the prime factorization, you have the information in a format that's very easy to compare. A number can be factored many different ways, but it can only be prime factored in one way. Once you have a list of the building blocks of two numbers, you can clearly identify the numbers they have in common. Combining those common building blocks will tell you the absolute biggest factor that those two numbers have in common. Here's a walk-through of how to find the greatest common factor again.

1. **Find the GCF of 7 and 24.** The prime factorization of 7 is just 7. The prime factorization of 24 is $2\times2\times2\times3$. Because these lists do not have any common numbers, you know that 7 and 24 do not have any prime factors in common. Thus, the GCF of 7 and 24 is 1. *When two numbers share no prime factors, their GCF is always 1.* That means that the fraction $\frac{7}{24}$ cannot be simplified.
2. **Find the GCF of 7 and 70.** The prime factorization of 7 is just 7. The prime factorization of 70 is $2\times5\times7$. The only prime factor that appears on both lists is 7, so 7 must be the GCF of 7 and 70. *When one number is a factor of another number, the smaller number will be the greatest common factor of the two.* Because 7 is a factor of 70 and can't have any factor higher than itself, 7 is greatest common factor of 7 and 70.
3. **Find the GCF of 220 and 18.** The prime factorization of 220 is $2\times2\times5\times11$. The prime factorization of 18 is $2\times3\times3$. The only number in common on those lists is one of the two 2s in the prime factorization of 220, so the GCF of 220 and 18 is 2. If you wanted to simplify the fraction $\frac{18}{220}$, the greatest factor you could cancel from both the numerator and denominator would be 2, which would leave you with $\frac{9}{110}$.

When you cancel out all the other factors, why are you left with 1?
Because "canceling out" the factors is really just dividing by them. So when you divide 2 by 2 to "cancel" these factors, they don't just disappear; they turn into 1. You don't usually bother writing the 1 down, because it doesn't have any effect. The only time you have to bother writing down 1 is when there are no other factors left.

Least Common Multiple

Finding the least common multiple of two numbers is another skill that will help you work with fractions. Being able to find the prime factorization of a number comes in handy when you are looking for the least common multiple of two numbers.

Find the Least Common Multiple

The **least common multiple** (or **LCM**) of two numbers is the smallest number that is a multiple of each number. For example, 6 is the least common multiple of 2 and 3. If you were to list out the multiples of 2, and then you were to list out the multiples of 3, 6 would be the smallest number to appear on both lists. Take a look:

Multiples of 2: 2, 4, 6, 8, 10, 12 . . .
Multiples of 3: 3, 6, 9, 12, 15 . . .

While there would be many numbers in common on these lists if you continued them indefinitely, the smallest, or "least," number on both lists is 6. Thus, 6 is the LCM of 2 and 3.

Much like with finding the greatest common factor, there is a much faster method of finding the least common multiple. And this method too relies on prime factorization.

Consider the numbers 24 and 144 again. This time, you want to find their LCM, so start by making a factor tree to find the prime factorization of each number.

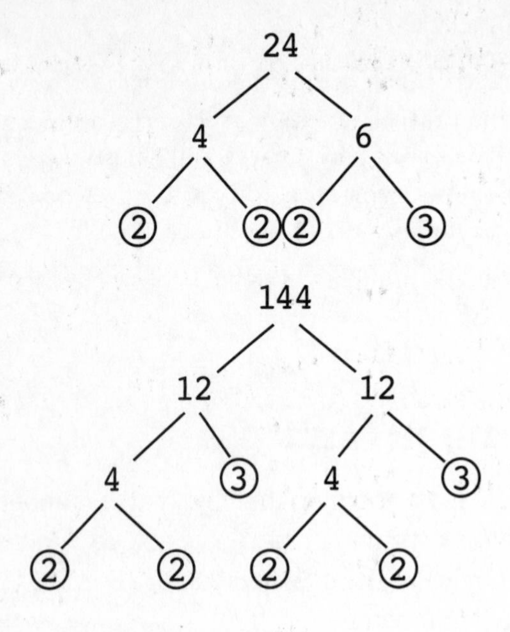

Now you know the prime factorization of each number, just as you did before.

Prime factors of 24: $2\times2\times2\times3$
Prime factors of 144: $2\times2\times2\times2\times3\times3$

Once again, look for the complete list of prime factors these numbers have in common. And once again, that list is $2\times2\times2\times3$. These factors are shared by 24 and 144, so they will show up in every multiple of 24, and in every multiple of 144. So you know they will be in your least common multiple, but you don't need them listed more than once.

A common multiple has to have *all* the prime factors on both lists. To be a multiple of each number, it has to have all the prime factors of each number. So once you've accounted for all the factors that overlap, you have to also account for all the remaining factors.

In this case, once you've accounted for the $2\times2\times2\times3$ that 24 and 144 have as common factors, you have no more factors in 24, but you still have a 2×3 remaining in 144. You need all those factors to show up in any number that will be a multiple of both 24 and 144. That means that the LCM of 24 and 144 is $2\times2\times2\times2\times3\times3$, which is 144.

If you were to list out all the multiples of 24, and then list out all the multiples of 144, 144 would be the smallest number to appear on both lists.

Steps for Finding the LCM of Two Numbers

1. Find the prime factorization of both numbers.
2. Find any common prime factors and list them *once*.
3. Add any other remaining, non-common factors from both numbers to the list.
4. Multiply the numbers in that list together to get your LCM.

Examples of Finding the LCM

1. **Find the LCM of 4 and 40.** The prime factorization of 4 is 2×2. The prime factorization of 40 is $2 \times 2 \times 2 \times 5$. The prime factors common to both numbers are 2×2. The remaining prime factors that are not common are 2×5. Thus, your LCM is $2 \times 2 \times 2 \times 5$, which equals 40. Just like in the last example, the LCM is the bigger of the two original numbers. This will happen any time you are trying to find the LCM of two numbers, and one of those numbers is a factor of the other.
2. **Find the LCM of 10 and 21.** The prime factorization of 10 is 2×5. The prime factorization of 21 is 3×7. There are no prime factors common to both numbers. The prime factors that are not common to both are $2 \times 5 \times 3 \times 7$. Thus, your LCM is $2 \times 5 \times 3 \times 7$, or 210. Notice that 210 is just the product of your two original numbers. This will happen any time you are trying to find the LCM of two numbers that have no prime factors in common.
3. **Find the LCM of 24 and 18.** The prime factorization of 24 is $2 \times 2 \times 2 \times 3$. The prime factorization of 18 is $2 \times 3 \times 3$. The prime factors common to both numbers are 2×3. The remaining primes that are not common to both numbers are 2×2 (the remaining factors of 24) and 3 (the remaining factor of 18). Thus, your LCM is $2 \times 3 \times 2 \times 2 \times 3$, or 72. Notice that 72 is bigger than both 24 and 18. This will happen any time there are non-overlapping factors in both of the numbers.

When you find the LCM of two numbers that have at least one prime factor in common, the LCM will be smaller than the product of the two numbers. Finding the LCM, much like finding the GCF, takes practice. It helps to get very comfortable with prime factorization, and it also helps to have a clear list of steps to follow, and to follow them carefully.

Chapter 4 Exercises

Identify each of the following numbers as "prime" or "not prime."

1. 2
2. 5
3. 15
4. 1
5. 21
6. 23
7. 83
8. 99

Write the following numbers in prime notation. (Note: You can use the multiplication sign if you prefer.)

1. 2
2. 14
3. 36
4. 18
5. 64

Find the greatest common factor (GCF) of the two numbers listed.

1. 14 and 21
2. 50 and 100
3. 17 and 40
4. 13 and 26
5. 90 and 18
6. 45 and 25
7. 44 and 66

Find the least common multiple (LCM) of the two numbers listed.

1. 13 and 2
2. 20 and 100
3. 18 and 24
4. 7 and 9
5. 9 and 36
6. 12 and 16

CHAPTER 5

Fractions

By clarifying the rules and process for how you write, alter, and manipulate fractions, you'll be getting ready to start adding variables into your math problems. The most important part of pre-algebra is clarifying the rules that you already know how to follow, so that you can still follow the rules when the numbers start getting replaced with variables. Fractions won't just show up by themselves anymore—they'll be mixed into math problems that test all different kinds of material.

Introduction to Fractions

There are three big ways that mathematicians represent non-integer numbers: decimals, fractions, and percentages. Let's say you have one orange, and you put a knife right in the middle and slice. You take a piece, and your friend takes a piece. You could say, "I have half of the orange." Or, your friend could say, "I have 50 percent of the orange." Or, you could say, "I have .5 oranges." All of these sentences mean the same thing.

Fractions are pretty much everyone's least favorite part of math, but they don't have to be. The key is to take the time to understand what they mean instead of just trying to memorize the rules you are supposed to follow. Once you understand *why* the rules are what they are, they are much easier to remember.

Numerator and Denominator

A fraction is just one number on top of another number. The top number is called the **numerator**, and the bottom number is called the **denominator**. These words are specific to fractions—they don't mean anything except "top number" and "bottom number."

The bottom number tells you how many pieces of something make up the whole. For example, when you cut that orange in half, there are two pieces, so the denominator would be 2. The top number tells you how many pieces you have. In the same example, the numerator would be 1, because you only have one piece of the orange.

You could look at the fraction $\frac{1}{2}$ and read it as, "Out of every two pieces, you have one." If you think about it for a second, that's exactly what the definition of the word *half* is. If you have a bag of candy, and you give your friend half, you will be giving her one out of every two pieces in the bag.

Most of the time when mathematicians talk about fractions, they are talking about fractions that are between 0 and 1. These fractions are called positive **proper fractions**. A proper fraction is a fraction whose numerator is smaller than its denominator; in other words, its absolute value is less than 1. Fractions that have a numerator bigger than their denominator are called **improper fractions**, and their absolute value is greater than 1. For example, $\frac{2}{3}$ is a proper fraction, but $\frac{3}{2}$ is an improper fraction.

Sometimes fractions are written with a horizontal line between the numerator and the denominator, like $\frac{2}{7}$ or $\frac{1}{3}$. Sometimes fractions are written with a slanted line between the numerator and the denominator, like 2/7 or 1/3. Whether you use a horizontal line or a slanted line does not have any effect on the size or meaning of the fraction; they are just two different ways of writing the same thing. The horizontal line makes it easier to add, subtract, multiply, and divide fractions, because the numerators and denominators will be matched up in straight lines.

Altering Fractions

Think again about the bag of candy discussed in the previous section. Let's say you want to give your friend one half of the candy, which is one out of every two pieces, or $\frac{1}{2}$. This means if there are forty pieces in the bag, you'd like to give her twenty; in that case you'd have twenty out of the forty pieces, or $\frac{20}{40}$. If there are six pieces in the bag, you both get three each, which means you end up with three pieces out of the six you had in total. Therefore, you get $\frac{3}{6}$ of the candy. If there are 100 pieces in the bag, you can keep fifty, so that's fifty out of 100, or $\frac{50}{100}$. This means that $\frac{1}{2}$, $\frac{3}{6}$, $\frac{20}{40}$, and $\frac{50}{100}$ must all be the same fraction, because they all give you the same relative information: no matter how many pieces of candy are in the bag, you can have half of them. Even though the actual number of pieces of candy is different in each situation, your friend is always eating half of your candy.

Infinite Possibilities

The same fraction can be written in an infinite amount of ways. You can keep changing the total number (the bottom number in a fraction) of pieces of candy in the bag, and that will change the number of pieces you give your friend (the top number in the fraction), even though you keep giving her half of your candy.

Look what happens: If you have two pieces of candy, your friend gets one. But if you have four pieces of candy, she gets two. If you double the candy, your share also doubles. If you have forty pieces of candy, your friend gets twenty. So when you multiplied the total pieces of candy by ten, your share also multiplied by ten. If you cut that number in half and go back to having only twenty pieces of candy in total, your share gets cut in half, too. That's the only way to keep the value of the fraction the same: to multiply or divide the top and bottom by the same number.

There are a few good reasons to alter fractions, or to write them in a different way. One reason is to simplify them so that they are easy to comprehend. Your friend probably wouldn't say, "May I please have five-tenths of your pizza?" if they could just say, "May I have half of your pizza?" So everyone naturally tries to simplify fractions by taking out any factors shared by both the top number and the bottom number.

The other big reason for altering fractions is to give them a common denominator. Remember, *denominator* is just a fancy word for whatever number is on the bottom of the fraction. (*Numerator* is the fancy word for the number on the top.) Whenever you need to add or subtract fractions, you must first make the denominators the same.

You conventionally give fraction answers in their most simplified form. A **simplified** fraction is one whose numerator and denominator share no common factors; in other words, the fraction cannot be reduced any further. Unless explicitly asked to do otherwise, you should simplify every fraction before you submit it as the answer to a problem on your homework, quiz, or exam.

Finding a Common Denominator

"Finding a common denominator" means altering two or more fractions with different bottom numbers so that the numbers on the bottom become the same, while the values of the fractions do not change. The most common reasons to do this are to compare the fractions, to add them, or to subtract them.

For example, let's say that you want to compare $\frac{1}{2}$ and $\frac{2}{3}$. If the denominators were the same, you'd only have to look at the numerators. **Once the denominators are the same, whichever fraction has the bigger numerator is the biggest fraction.** So you will need to change $\frac{1}{2}$ and $\frac{2}{3}$ so that the denominators are the same. To do this, you can use any number that has both 2 and 3 as factors. The number 6 comes to mind as an easy choice. To get started, change both fractions so that they have a 6 on the bottom. How do you do that? By multiplying both the top and bottom of each fraction by whatever number makes the bottom number 6.

Take a look at $\frac{1}{2}$. What would you multiply the denominator by to get 6? You'd multiply it by 3. Therefore, because you don't want the value of the fraction to change, you have to multiply both the numerator and the denominator by 3.

$$\frac{1}{2} \times \frac{3}{3}$$

Since $\frac{3}{3} = 1$, you're just multiplying by 1, so you're not changing the value of the fraction. You're just able to express it in a more useful way.

$$\frac{1}{2} \times \frac{3}{3} = \frac{(1 \times 3)}{(2 \times 3)} = \frac{3}{6}$$

Giving your friend one out of every two pieces of pizza is the same as giving him three out of every six pieces.

Now, do the same thing to $\frac{2}{3}$. You need the denominator to be 6, and right now it's 3, so that means you have to multiply by 2. Because you don't want to change the value of the fraction, you have to multiply the numerator and denominator by the same number.

$$\frac{2}{3} \times \frac{2}{2} = \frac{(2 \times 2)}{(3 \times 2)} = \frac{4}{6}$$

Now that these fractions have the same denominator, you are able to do three things easily:

1. **Compare them.** Now that you know one of the fractions is $\frac{3}{6}$ and the other is $\frac{4}{6}$, it's clear which fraction is bigger. The first fraction gives you three out of every six pieces, and the second fraction gives you four out of every six pieces. Thus, $\frac{2}{3}$ is bigger than $\frac{1}{2}$.
2. **Add them.** When you add or subtract fractions, the first step is to get a common denominator. Then, to add the fractions, you just add the numerators and leave the denominators alone. You can see that when you add $\frac{3}{6}$ and $\frac{4}{6}$, you have a total of $\frac{7}{6}$.
3. **Subtract them.** Subtracting fractions is a lot like adding them. You find a common denominator, keep it, and just subtract the numerators. Thus $\frac{4}{6}-\frac{3}{6}=\frac{1}{6}$. Subtraction finds the difference between two numbers, and the difference between "four out of every six pieces" and "three out of every six pieces" is "one out of every six pieces."

While you *can* always find a common denominator by multiplying the two different denominators together, that's not always the easiest move. For example, think about $\frac{1}{5}$ and $\frac{10}{50}$. To compare, add, or subtract them, you first need to find a common denominator. It's true that you could multiply $5\times50=250$, and use that as your common denominator, but do you see a much smaller number you can use? How about 50 itself?

You would end up with the same answer either way, but the first way forces you to do a lot of extra multiplying and then a lot of extra simplifying, which is a lot of extra work.

In order to find the smallest possible common denominator, and thus avoid a lot of extra work, you are going to look for the least common multiple of the denominators you want to make common.

Adding and Subtracting Fractions

Manipulating fractions is difficult for many people, so don't feel discouraged if you are having problems in this arena. Lawyers, investment bankers, and even accountants have been brought almost to tears when asked to manipulate fractions. That's because they tried to memorize the rules for manipulating fractions without actually understanding them.

You do have to remember the rules, but that isn't enough. If you understand where the rules come from, you're much more likely to remember how to use them. The rules for adding, subtracting, multiplying, and dividing fractions are just shortcuts to help you solve for what must be true in the real world.

Real-World Fractions

When you add fractions in the real world, they don't have to have a common denominator. You can put $\frac{1}{3}$ of a pizza and $\frac{1}{5}$ of a pizza in a box together. There: you've added them. But how much of the pizza do you have in the box now? That's where you need a common denominator.

The denominator is the unit of measure, and the numerator tells you how many of that unit you have. When you say $\frac{1}{3}$, you literally say "one third." That's like saying "one orange" or "one apple." So when you try to add one "third" to one "fifth," you can't put them together. It's like adding apples and oranges. All you can say is "I have one apple and one orange."

But if you were to find some common way of describing these units, then you could add them with no problem. For example, an apple and an orange are both pieces of fruit. So if they are described in that way, you are now adding one piece of fruit to another piece of fruit, and the final result is two pieces of fruit.

This is exactly what a common denominator does. Just like describing both an apple and an orange as pieces of fruit, it gives you a common unit description of two fractions so that you can add them together. It doesn't change the value of the numbers: an apple is still an apple, and an orange is still an orange, even if you refer to them each as a piece of fruit.

Examples

For example, if you want to add $\frac{1}{3}$ and $\frac{1}{5}$, you'll need to find the easiest way to describe these two fractions in terms of a common denominator. The smallest common denominator of two fractions will be the least common multiple of the two denominators. Think about what "least common multiple" means: it means the smallest number that is a multiple of both numbers.

Therefore, the LCM of 3 and 5 is 15. That means you want to convert the denominators of each fraction into 15 without changing the actual size of the fraction, so you'll need to convert $\frac{1}{3}$ into something with a 15 on the bottom. Since you're multiplying the denominator by 5, you also need to do the same thing to the numerator. This won't change the size of the fraction, because you're multiplying it by $\frac{5}{5}$, which is the same as multiplying by 1. So instead of $\frac{1}{3}$, you can say $\frac{5}{15}$. It's exactly the same. Just as you can reduce $\frac{5}{15}$ to $\frac{1}{3}$ by factoring out the 5 that the numerator and denominator have in common, you can "inflate" $\frac{1}{3}$ by multiplying both the numerator and the denominator by the same factor.

Using the same process, you can "inflate" $\frac{1}{5}$ to $\frac{3}{15}$ by multiplying the numerator and the denominator by 3. And now instead of adding $\frac{1}{3}$ and $\frac{1}{5}$, you're adding $\frac{5}{15}$ and $\frac{3}{15}$. You are no longer adding apples and oranges; you're adding fruit and fruit. After "inflating" the fractions, fifteenths are the units. You have three fifteenths and five fifteenths, which means, in total, you have eight fifteenths, or $\frac{8}{15}$.

That's where the rules about addition of fractions come from. You need a common denominator in order for the units to be the same. Once the units are the same, you leave the denominators as they are and simply add the numerators.

Subtracting fractions works exactly the same way, for exactly the same reasons. You can't calculate one apple minus one orange. There's no way to put that information together. But you can calculate one piece of fruit minus one piece of fruit. So you need to get the units, or the denominators, the same before you can subtract. Once the denominators are the same, you simply subtract the numerators. For example, $\frac{1}{3}-\frac{1}{5}$ would convert to $\frac{5}{15}-\frac{3}{15}$, which gives you $\frac{2}{15}$. Just like "five pieces of fruit minus three pieces of fruit equals two pieces of fruit," it's also true that "five fifteenths minus three fifteenths equals two fifteenths." Once you change the denominator, you'll know what your units of measure are, and then you can add or subtract just as you would with any other units of measure.

Multiplying and Dividing Fractions

Multiplying and dividing fractions is actually a whole lot easier than adding or subtracting them. That's because multiplying and dividing fractions does *not* require a common denominator.

When mathematicians talk about multiplication, they almost always use the word *by*. They say things like "multiply 2 by 5" or "When 2 is multiplied by 5, the product is 10." But you can also think about multiplication in terms of how many *of* something you have. When you want to talk about the expression 2×3, you could just as easily say that you want to find two groups of three. So, when you multiply some number by 2, you're finding two of that number. When you multiply something by 10, you're finding ten of that number. When you multiply something by 1, you're finding one of that number, which is why the final value isn't altered at all when something is multiplied by 1.

Multiplying Fractions

You can think of multiplying by a fraction in the same way. When you multiply any number by $\frac{1}{2}$, you're finding half of that number. You know that $\frac{1}{2}$ of 4 is 2, or $\frac{1}{4}$ of 100 is 25. That *of* is exactly the same as saying "multiplied by." So you can think of these equations as "$\frac{1}{2}$ multiplied by 4 is 2" or "$\frac{1}{4}$ multiplied by 100 is 25." When you multiply a positive number by a proper fraction, you're actually taking a fraction of that number, and thus making it smaller.

When you multiply a fraction by an integer, it helps to make both numbers into fractions. You do this, of course, by putting the integer over the number 1. So let's say you're asked to find:

$$\frac{1}{2} \text{ of } 80$$

You know that finding a fraction of a number is the same as multiplying the fraction by that number, so you can rewrite the expression as:

$$\frac{1}{2} \times 80$$

Now that you're multiplying a fraction by an integer, you want both numbers in fraction form. To turn an integer into a fraction, put a 1 in the denominator:

$$\frac{1}{2} \times \frac{80}{1}$$

What's the rule for multiplying fractions? Multiply the numerators to get your new numerator. Multiply the denominators to get your new denominator.

$$\frac{1}{2} \times \frac{80}{1} = \frac{(1 \times 80)}{(2 \times 1)} = \frac{80}{2}$$

Now all you have left to do is simplify: 80 and 2 have 2 as their greatest common factor, so you can factor a 2 out of both the numerator and denominator:

$$\frac{80}{2} = \frac{(80 \div 2)}{(2 \div 2)} = \frac{40}{1}$$

Since any number divided by 1 is itself, $\frac{40}{1}$ is equal to 40.

You probably already knew that $\frac{1}{2}$ of 80 is 40 without all these steps, but it's good proof that the rules work. When you're multiplying fractions, you multiply the numerators, multiply the denominators, and then simplify the result. That's it!

In essence, fractions are a way of representing multiplication and division at the same time, which might be why people find them so tricky. The denominator tells you how many pieces you are breaking the number into: that's the division part. The numerator then tells you how many of those pieces you have: that's the multiplication part. So, when you multiply something by $\frac{1}{2}$, you're really dividing it by 2 (breaking it into two parts) and then multiplying it by 1 (taking one of those parts). That's why, when you took $\frac{1}{2}$ of 80, you ended up with the same answer you would have found if you had taken 80 and divided it by 2.

Don't Forget to Simplify

One more little tip about multiplying fractions: if you can simplify along the way, you'll make your life a lot easier! For example, let's say you've been asked to find:

$$\frac{15}{7} \times \frac{42}{5}$$

You already have both the numbers in fraction form, so you know that you're supposed to multiply the numerators, multiply the denominators, and then simplify. Taking those steps would indeed lead to an answer, but you'd have to do a lot of work to get there. Here's your equation:

$$\frac{15}{7} \times \frac{42}{5} = \frac{(15 \times 42)}{(7 \times 5)}$$

You know that $7 \times 5 = 35$, but you will have to do 15×42 out on paper to figure out that it was equal to 630. Once you do that, you'll have $\frac{630}{35}$. You know that you must be able to simplify, because you can see that both the numerator and the denominator are divisible by 5. So you factor a 5 out of both the numerator and the denominator:

$$\frac{630}{35} = \frac{(630 \div 5)}{(35 \div 5)}$$

Again, you probably don't know these values at a glance. After a little long division, though, you'll be left with $\frac{126}{7}$. Now you have to figure out whether 7 is a factor of 126, and since there's no divisibility trick for 7, you're going to have to do out some long division to see if 126 is divisible by 7. It is! Now you're able to simplify further.

$$\frac{126}{7} = \frac{(126 \div 7)}{(7 \div 7)} = \frac{18}{1} = 18$$

Yikes, that was a lot of work! How could you make your life easier? You could have simplified *before* you did the multiplication. Let's go back a few steps to where you multiplied the numerators and denominators to get:

$$\frac{15}{7} \times \frac{42}{5} = \frac{(15 \times 42)}{(7 \times 5)}$$

You can clearly see that the numerator and denominator both have 5 and 7 as factors. You don't need to do the multiplication here just to undo it later. So you can, right now, factor a 5 and 7 out of the numerator and the denominator. And what do you have left?

$$\frac{(15 \times 42)}{(7 \times 5)} = \frac{(15 \times 42 \div 7)}{(7 \times 5 \div 7)} = \frac{15 \times 6}{5} = \frac{(15 \times 6 \div 5)}{(5 \div 5)} = \frac{(3 \times 6)}{1} = \frac{18}{1} = 18$$

You still end up with 18 as the answer, but you have saved yourself a lot of trouble along the way.

Dividing Fractions

Dividing fractions isn't any harder than multiplying fractions, but it has a weird rule that can be hard for people to remember. The rule is that dividing by a number is the same as multiplying by the reciprocal of that number. The **reciprocal** of a number is what you get when you "flip it over," switching the numerator and the denominator. For example, $\frac{2}{3}$ is the reciprocal of $\frac{3}{2}$, and $\frac{1}{2}$ is the reciprocal of $\frac{2}{1}$, or 2.

Keep in mind that multiplying a number by $\frac{1}{2}$ is the same as taking $\frac{1}{2}$ of that number, which is the same as dividing the number by 2. See? Multiplying by $\frac{1}{2}$ is the same as dividing by 2, because multiplying by a number is the same as dividing by its reciprocal.

What happens when you divide a number by a positive proper fraction? You actually make the magnitude of the number bigger. Thus when you divide a number by $\frac{1}{2}$, you're actually multiplying the number by the reciprocal of $\frac{1}{2}$, which is 2. That can be a difficult concept to wrap your head around, because you never say, "I want to divide this *by* half." Most people divide things *in* half. If you divide something *in* half, that's the same as taking half of it, or multiplying it by $\frac{1}{2}$.

But when you divide something *by* a number, you break it into pieces. When you divide something by 4, for example, you're basically saying, "If I break this number evenly into four groups, how many are in each group?" The answer will be $\frac{1}{4}$ the size of the original number, because dividing by 4 is the same as multiplying by $\frac{1}{4}$. When you divide something by $\frac{1}{2}$, you are basically saying, "If I make this number into half a group, how many would be in the whole group?" If a number is half of the group, the whole group must be 2 times that number. That's why dividing by $\frac{1}{2}$ is the same as multiplying by 2.

Dividing by a number is the same as multiplying by its reciprocal. Multiplying by a number is the same as dividing by its reciprocal.

This is what makes dividing by a fraction easy to learn: you never have to do it. Instead of dividing by a fraction, you multiply by its reciprocal. Take the following example:

What is $\frac{2}{3} \div \frac{3}{4}$?

You know that dividing by a fraction (or by any number) is the same as multiplying by its reciprocal; so first, you should rewrite the question as:

What is $\frac{2}{3} \times \frac{4}{3}$?

Now you just have to multiply the numerators, multiply the denominators, and simplify. Once you multiply, you have:

$$\frac{2}{3} \times \frac{4}{3} = \frac{(2 \times 4)}{(3 \times 3)} = \frac{8}{9}$$

When you divide a positive number by a proper fraction, it gets bigger. That should be the common-sense check when you divide a number by a positive proper fraction. The number should get bigger because dividing by a fraction is the same as multiplying by its reciprocal.

Mixed Numbers

Mixed numbers are numbers that are written to include both an integer and a proper fraction. For example, $2\frac{1}{2}$ and $3\frac{3}{4}$ are mixed numbers. You would read these mixed numbers as "two and a half" or "three and three quarters." Mixed numbers are really useful for verbally communicating information and explaining things in words. For example, you might say that you need "one-and-a-half cups of sugar" for a recipe, or that you have "three-and-a-half hours" before you have to be somewhere.

But mixed numbers really aren't very useful for doing math. Sure, sometimes they are numbers you are familiar with, and then you can do the math in your head. You can probably figure out that if you add two-and-a-half cups of water to two-and-a-half cups of milk, you will have five cups of liquid (and it will taste gross). But what about if you add $3\frac{5}{8}$ cups of sugar to $4\frac{2}{5}$ cups of flour? What if you have $4\frac{2}{5}$ cups of flour and you take $3\frac{5}{8}$ cups away? Now the mixed numbers are trouble. Your plan of attack is to convert mixed numbers to fractions if you have to do math with them.

Converting Mixed Numbers to Fractions

Generally, when faced with mixed numbers, you'll find it easiest to convert them into fractions. You already know how to add, subtract, multiply, and divide fractions, so by converting them, you're creating math that you know how to do. And that's always a good move! *How* do you convert a mixed number into a fraction? Follow these steps.

Let's use $3\frac{5}{8}$ as an example. If you were to read this number, you'd read it as "three and five eighths." So another way to express $3\frac{5}{8}$ is to write "3 and $\frac{5}{8}$," which is the same as "$3+\frac{5}{8}$." Now you have to add an integer and a fraction, which you already know how to do: turn the integer into a fraction, give the two fractions common denominators, and then add the numerators. Check it out:

$$3+\frac{5}{8}=\frac{3}{1}+\frac{5}{8}=\frac{(3\times8)}{(1\times8)}+\frac{5}{8}=\frac{24}{8}+\frac{5}{8}=\frac{29}{8}$$

> Statements like the one above are going to be confusing if you try to look at them all at once. When you see a long statement in this book or in your textbook, just work one equal sign at a time. Every time you hit another equal sign, you get a new piece of information.

You can see why it's difficult to *talk* about mixed numbers this way. If you told your friend, "I need $\frac{29}{8}$ cups of sugar," he would probably look at you a little funny. But $\frac{29}{8}$ will be much easier to add, subtract, multiply, or divide than the mixed number $3\frac{5}{8}$.

Now here's a shortcut for converting a mixed number to a fraction, using the same example of the mixed number $3\frac{5}{8}$.

1. Multiply the integer by the denominator of the fraction. Here, the integer is 3, and the denominator of the fraction is 8. So you would multiply them: $3 \times 8 = 24$.
2. Take the product from Step 1 and add it to the numerator of the fraction. $24 + 5 = 29$.
3. The sum from Step 2 is your numerator. Keep the original denominator and write your new fraction. In this example, 29 is your new numerator and 8 is your old denominator, so you would write $\frac{29}{8}$.

In this example, by multiplying 3 × 8, you are trying to find the number you could divide by 8 (by having an 8 in the denominator) that would result in a value of 3. That number, of course, is 24, because 24 divided by 8, or $\frac{24}{8}$, is equal to 3.

Converting Fractions to Mixed Numbers

Sometimes, you may be asked to convert your fraction into the form of a mixed number. Luckily, to accomplish this, you just have to do the exact opposite of what you did to convert a mixed number to a fraction! Let's take a look.

Let's say that you finish a fraction problem and get the solution of $\frac{27}{6}$. If the directions ask you to convert your answer to a mixed number, determine whether the fraction can be converted to a mixed number (you are checking to see if the fraction is improper: the numerator is bigger than the denominator). Since the fraction is bigger than one, it *can* be converted to a mixed number.

Then take a look at the denominator, which is 6. You know that when you're done, the numerator will be smaller than 6. Basically, every time there is a full group of 6 in the numerator, you want to pull it out of the fraction, because $\frac{6}{6}$ is really 1, which can be written as an integer.

Next, ask yourself, "How many sixes are in the numerator?" Here, you are trying to find the biggest multiple of 6 that's smaller than 27. Six goes into 27 only four times. Therefore, the integer for this mixed number is 4.

What's left over when you take out that 4? Well, removing 4 from the fraction is the same as removing $\frac{24}{6}$ from the fraction, since $\frac{24}{6}$ is equal to 4. So you need to subtract $\frac{24}{6}$ from the $\frac{27}{6}$ you started with, and you know that this just means subtracting the numerators. Now you are left with $\frac{3}{6}$ as your fraction. So it must be true that $\frac{27}{6}$ is equal to $4\frac{3}{6}$.

But wait! You aren't quite finished yet, because you always want to write your answer to fraction problems in the most reduced form possible. You know that $\frac{3}{6}$ is the same as $\frac{1}{2}$. So reduce the fraction, and now you're done! The mixed number way to write $\frac{27}{6}$ is $4\frac{1}{2}$.

Notice that what you've done to convert a fraction to a mixed number is the reverse of what you did to convert a mixed number to a fraction. To make it easy, here are the steps to convert $\frac{27}{6}$ to a mixed number:

1. Divide the numerator by the denominator as many times as it fits evenly: 6 goes into 27 four full times. So you have four groups of 6, or 24 sixths, to pull out of the numerator.
2. Subtract the value from Step 1 from the numerator. In this example, $27-24=3$, so 3 is your remaining numerator, and it's still over the denominator of 6. So the fraction part of your mixed number will be $\frac{3}{6}$.
3. Convert the value from Step 1 into an integer by dividing it by the denominator: $24\div 6=4$.
4. Combine your integer and fraction. Step 2 tells you the fraction piece of the mixed number is $\frac{3}{6}$, and Step 3 tells you the integer piece of the mixed number is 4. So the mixed number is $4\frac{3}{6}$.

Of course you can't leave $4\frac{3}{6}$ as the answer; you should simplify this answer to $4\frac{1}{2}$. Keep in mind that you wouldn't have had to reduce the fraction at the end (and would have made your life a lot easier along the way) if you'd reduced $\frac{27}{6}$ when you first started working with it. Like this:

$$\frac{27}{6}=\frac{27\div 3}{6\div 3}=\frac{9}{2}$$

You can see that 2 goes into 9 four times, so you have a 4 to pull out as your integer. And when you pull out that 4, you're left with $\frac{1}{2}$. You still get $4\frac{1}{2}$ as an answer; you just get there with a lot less work because the numbers are smaller and easier to work with!

Ratios

Ratios are a special kind of fraction. A **ratio** is a relationship between two amounts that shows the number of times one value contains another. Well, that's the official definition, but it's difficult to understand. In simpler terms, a ratio shows the size relationship of two different amounts. A ratio doesn't give you the actual amounts involved; it just tells you how big one amount is in comparison to another amount.

Instead of talking in circles about the definition of ratios, let's look at an example. What if someone tells you to take some juice concentrate and mix it with water so that "the ratio of concentrate to water is one to three"? What does that mean?

It means that for every one part concentrate you dump into a pitcher, you have to dump three parts of water in there. That way, the mixture you create will be one part concentrate for every three parts water, which means that the concentrate and water are in a ratio of one to three.

This ratio doesn't tell you how much juice you have. That "one part" concentrate could be one cup, mixed with three cups of water. Or that "one part" concentrate could be ten gallons of concentrate, in which case you would have to put thirty gallons of water in your very, very big pitcher to keep the ratio correct.

That's how ratios work. For example, the following three statements all convey exactly the same information.

- The ratio of concentrate to water is 1 to 3
- The ratio of concentrate to water is $1:3$
- The ratio of concentrate to water is $\frac{1}{3}$

This information can be given in words (such as in the first statement); it can be given with a colon used as the ratio symbol (such as in the second statement); or it can be given with a fraction bar as the ratio symbol (such as in the third statement). Your teacher will probably prefer you to use either the colon or the fraction bar to represent a ratio. The fraction bar is especially important to remember, because it will help remind you that ratios can be altered in all the same ways that fractions can be altered.

For example, if someone tells you that the ratio of sugar to butter in a certain cookie recipe is $1:2$, that's no different than having a sugar to butter ratio of $2:4$. That's because $\frac{1}{2}$ and $\frac{2}{4}$ are the same fraction. Remember, a ratio doesn't tell you how much sugar or butter you have; it just tells you how the amounts of sugar to butter compare to one another. So you can reduce a ratio the same way you simplify a fraction: by dividing the numerator and denominator by any factor they share.

Chapter 5 Exercises

Simplify the following expressions.

1. $\frac{1}{2}+\frac{2}{3}$
2. $\frac{1}{3}+\frac{2}{9}$
3. $\frac{3}{5}-\frac{2}{10}$
4. $\frac{1}{2}-\frac{1}{2}$
5. $\frac{2}{9}\times\frac{9}{2}$
6. $\frac{1}{3}\times\frac{9}{8}$
7. $\frac{1}{2}\div\frac{1}{2}$
8. $\frac{1}{4}\div\frac{1}{2}$

Convert the following fractions to mixed numbers.

1. $\frac{9}{3}$
2. $\frac{13}{4}$
3. $\frac{7}{2}$

Convert the following mixed numbers to fractions.

1. $3\frac{1}{3}$
2. $7\frac{1}{2}$
3. $1\frac{2}{9}$

CHAPTER 6

Decimals and Percent

Although the mechanics are slightly different, decimals and percentages are just two more ways of expressing fraction-like information. Most of the time in pre-algebra and beyond, you're going to find yourself working with fractions instead of decimals and percentages. But you have to be able to convert information from one form to another so that you can understand it, use it, and give your answers in the form required by your homework, quizzes, or tests.

Introduction to Percent

Percentage is particularly good for comparing lots of numbers in a way that is meaningful to people. You will be using percent throughout your entire life. The word root *cent* means "one hundred." Centipedes look like they have 100 legs; a centimeter is a hundredth of a meter; and there are 100 cents in a dollar. So the word *percent* literally means "per cent" or "per 100 parts." That's what percent is: when you say "fifty percent," you're saying "fifty out of every 100." That's why 50% is the same as $\frac{1}{2}$: because 50% means $\frac{50}{100}$, which reduces to $\frac{1}{2}$.

100% is the total of something, because it is 100 out of every 100 parts. If you talk about a percent bigger than 100%, you're talking about more than the whole thing. So while you can't spend 110% of your savings or "give 110%," you *can* have a bank account that increased 200% last year (because it now contains three times as much as it did before).

While there are negative decimals and fractions, there is no negative percent. That's because percent is a little different than decimals and fractions, in that it always has to be a percent *of* something. You can't just have 40%. As an amount, it doesn't mean anything. You'd have to say "40% of the candy," or "40% of your money," or something like that. You can do the same thing with decimals and fractions: "$\frac{2}{5}$ of your candy" and ".4 of your money" have the same meaning as their percent counterpart. But the numbers $\frac{2}{5}$ and .4 also mean something when they stand on their own. $\frac{2}{5}$ is a number: it is between 0 and 1 on the number line. It's in the exact same place as the number .4, because it has the same value. Forty percent isn't a number. You can't put 40% on the number line. Because of this, percent can't be negative. Instead, you can use phrases like "a 100% decrease" or "50% less than," since words such as "decrease" and "less than" show that the number is getting smaller.

Often, percent questions include rounding. When you are asked to "round to the nearest percent," your percent should be a whole integer. So for example, if you were asked to round 53.46% to the nearest percent, you would rewrite it as 53%. If you were asked to round 6.6% to the nearest percent, you would rewrite it as 7%.

Introduction to Decimals

Most people are somewhat comfortable with decimals because they have dealt with money. Think about the value $10.45, which stands for ten dollars and forty-five cents. The value represented by the numbers to the right of the decimal point is the part that is less than a full integer; it's a piece of an integer. That's why $10 is the same in value as $10.00. The numbers to the right of the decimal place show you pieces of a dollar, and in both these cases, there are none. In the case of $10.45, you can think of forty-five cents as forty-five parts of a dollar, or forty-five of 100 cents. That .45 is, for this reason, the same as $\frac{45}{100}$.

Every integer could be written with a decimal point. The number 10, for example, could be written as 10.0 or 10.00 or 10.0000000, and these all have exactly the same value. No matter how many zeros you put to the right of the decimal place, there are still no pieces there.

Even if you don't see the decimal point, it's still there. Every number can be written with a decimal point, placed after the units digit. Everything to the left of it is an integer. Everything to the right of it is a piece of an integer.

Changing the Value

While some of the zeroes to the right of the decimal point don't matter, some of them do. It depends on where you insert them. Take a look at the number 10.4 as an example. If you change the number to 10.04, you change its value. Think in terms of money: 10.4 is the same as $10.40. Is that the same as $10.04? It's not. So you can stick all the zeroes to the right of the number that you want, but if you insert zeroes anywhere between the decimal point and a non-zero digit, in a way that moves a non-zero digit or digits to a different place value, you're changing the value of the number. The number 10.4 is bigger than the number 10.04, which is bigger than 10.0004, which is bigger than 10.000004. So you can't just put zeroes wherever you want, because you'll change the value of the number.

You *can* put all the zeroes you want after the values that are given to you. You can take 10.4 and write it as 10.40 or 10.4000 or 10.4000000000, and it won't matter! That 4 is staying exactly where it is, thus preserving its value. You can tack all the zeroes you want on the end without changing the value of the number, because zeroes have no value at all.

Manipulating Decimals

You can add and subtract decimals just like any other number. The only rule to remember is that you must line up the decimal points. That shouldn't surprise you—any time you add up numbers, you align the place values. So if you're adding $3.5+2.7$, you line up the decimal points and add as normal, just like you were adding $35+27$, but in the end, you leave the decimal point where it is. So $3.5+2.7=6.2$. To make things a little trickier, what if you have to add 4.5 to 2.0003? To add any decimals together:

1. Line up the decimal points. In this example, you'd line up the numbers like this:

   4.5
   2.0003

2. If one decimal is longer than the other, add zeroes to the right of the shorter decimal:

   4.5000
   2.0003

3. Add as normal, keeping the decimal wherever it is.

   6.5003

Subtraction works exactly the same way. In short, adding numbers and subtracting numbers doesn't change at all when there are decimals involved.

Multiplication and division of decimals are easy on the calculator. Multiplication and division with decimals on paper are a tiny bit trickier than addition and subtraction, but there's really only one extra step you have to remember. For example, how would you multiply 25×3.38 on paper?

1. Ignore the decimal point and multiply the two numbers as if it weren't there at all. So in this example, you would do $25 \times 338 = 8,450$.
2. Count up the total number of digits to the right of the decimal points in the numbers you multiplied. In this case, there are two in all: two digits to the right of the decimal point in 3.38 and no digits to the right of the decimal point in 25, for a total of two digits to the right of the decimal points in the original numbers.

For the second step, do *not* count any zeroes to the right of a decimal point that don't have any other values after them. For example, you could have written 25 as 25.000, but those zeroes are nonessential and don't really do anything, so you don't count them. However, if you had the number 25.005, those zeroes count. The number 25.0050 has three digits to the right of the decimal point.

3. Place a decimal point in your product so that there is the same number of digits to the right of the decimal point in your product as the number you found in Step 2. Thus, you take your product of 8,450 from Step 1 and place the decimal point two digits in from the right to get 84.50.

Dividing Decimals

Dividing with decimals has a similar process. Take a look at these three examples to see how it works.

- **Possibility one:** The dividend has a decimal, and the divisor doesn't. (In a division problem, the dividend is the first number and the divisor is the second.) So, for example, if you were asked to find $\frac{25.2}{18}$, you would start by just doing $252 \div 18 = 14$. Then, you do what you would do in a multiplication problem: count up the digits to the right of the decimal point in the dividend and put the decimal point in the quotient so that there is the same amount of digits to the right of the decimal place. Now you know $25.2 \div 18 = 1.4$. Why? Because when you changed 25.2 into 252, you made it 10 times bigger. To get the right answer, you need to undo this change by making the result ten times smaller, or dividing by 10.

- **Possibility two:** The dividend doesn't have a decimal, but the divisor does. So, for example, you're asked to simplify $25 \div 1.25$ to be written as one number. First, you can ignore the decimal point and just divide $25 \div 125$, which will give you .2 as an answer. Now, you have to count up the digits to the right of the decimal place in the divisor: in this problem, there are two. So, you've made the divisor 100 times bigger than it should be by moving the decimal point two places to the right. You have to compensate for that by multiplying the product you found by 100. You've got to move the decimal place two places to the right: that means you change .2 to 20. Now you have $25 \div 1.25 = 20$. Basically, you moved the decimal point to the right two places in the divisor when you started the problem, so you have to move the decimal point to the right two places in the quotient when you finish the problem.
- **Possibility three:** The dividend and the divisor both have decimals. In this case, you combine the rules above. Say you are dividing $2.5 \div 1.25$, and you're trying to simplify to one number. Dividing 25 by 125 gives you .2 as an answer. Your dividend of 2.5 has one digit after the decimal point, so you have to move the decimal point in your quotient one place to the left. But your divisor of 1.25 has two digits to the right of the decimal point, so you have to move the decimal point in the quotient two places to the right. Moving the decimal point one place to the left and two places to the right results in the decimal point moving one place to the right, which changes your answer to 2.

Converting among Decimals, Fractions, and Percent

Any one of these representations of a value (decimal, fraction, or percent) can be converted to the other. Keep in mind that the goal of this conversion is to preserve the *value* of the number while just writing it in a different way.

Converting a Fraction to a Decimal

The easiest way to convert a fraction into a decimal is with long division. Remember that a fraction is really just a representation of division: the fraction bar tells you to take the numerator and divide it by the denominator.

So if you're asked to convert $\frac{1}{2}$ into a decimal, for example, you just want to take 1 and divide it by 2. Since 2 doesn't go into 1 at all, you're forced to add a decimal place to the long division, and then you can figure out that 2 goes into 1.0 a total of .5 times. Then you have your decimal conversion!

Converting Mixed Numbers to Decimals

Sometimes, you may be asked to convert mixed numbers into decimals. Let's go back to the example of $3\frac{5}{8}$ to see how it's done. Follow these steps:

1. Leave the integer alone. Stick it to the left of the decimal point and be done with it.
2. Convert the fraction part of the mixed number into a decimal. For this, you have to change $\frac{5}{8}$ into decimal form. Find the decimal by dividing the numerator by the denominator. Remember, sometimes a fraction won't convert evenly to a decimal, and you will have to round your answer or express it as a repeating decimal. In this example, you have to convert $\frac{5}{8}$ to a decimal. Dividing 8 by 5 gives you .625.
3. Put the integer and decimal back together. Doing so reveals the decimal conversion of $3\frac{5}{8}$ is 3.625.

Converting a Fraction to a Percent

Converting a fraction to a percent isn't something you will do that often. Most of the time, it's easier to work with fractions and decimals. You really only convert fractions to percent when a problem explicitly asks you to give the answer as a percent. If you have to convert a fraction to a percent, how do you do it? As explained previously, percent literally means "per 100," so in order to convert a fraction to a percent, the first step is to make the denominator into 100.

Sometimes this will be pretty easy. Take $\frac{1}{2}$, for example. How do you convert the denominator into 100? You multiply it by 50. That means you have to multiply the numerator and denominator by 50, which gives you $\frac{50}{100}$. Now that you know the fraction $\frac{1}{2}$ can be represented as $\frac{50}{100}$, you've converted the fraction to 50%.

To change a fraction into a percent, alter the fraction so that the denominator is 100. Once you've done that, the numerator of your new fraction is your percent. Why? Because a percent tells you how many parts out of 100, and a numerator over 100 tells you the same thing.

Sometimes the conversion might be a little messier. Think about the fraction $\frac{7}{30}$. If you were asked to represent that fraction as a percent, you'd be stumped at first, because there's no integer you can multiply by 30 to yield 100 as a denominator. You're going to have to start by multiplying the numerator and the denominator by 10, which gives you $\frac{70}{300}$. Now you can see how to convert the denominator to 100: you have to divide it by 3. And since you have to do the same thing to the numerator that you do to the denominator, you have to divide them both by 3. The denominator is now 100. Mission accomplished. What happens to the numerator? Nothing pretty:

$$\frac{70}{300} = \frac{(70 \div 3)}{(300 \div 3)} = \frac{(70 \div 3)}{100}$$

The denominator is 100, so the numerator must be your percent. In this case, the percent is $\frac{70}{3}$. If you wrote it as a mixed number, it would be $23\frac{1}{3}\%$. If you were asked to round your answer to the nearest percent, it would be 23%.

Converting a Percent to a Fraction

Converting a percent into a fraction is a lot easier than converting a fraction into a percent. That's because percent, by definition, gives you the number of pieces out of every 100. So you can take your percent and put it over 100, and then you've got a fraction.

That's the whole rule: Put the percent number (ignore the percent sign) over 100, and reduce.

So if you're asked, for example, to convert 8% into a fraction, you just take 8 and put it over 100. So $8\% = \frac{8}{100}$. The numerator and denominator share a common factor of 4, so you can reduce the fraction. $8\% = \frac{8}{100} = \frac{8 \div 4}{100 \div 4} = \frac{2}{25}$.

Converting a Percent to a Decimal

Let's say that you are asked to convert 40% into a decimal. You know how to convert a percent to a fraction: drop the percent sign and put the percent amount in the numerator and the number 100 in the denominator. So you have $\frac{40}{100}$. Now you just have a division problem, which you can do to get a decimal. When you divide a number by 100, you move the decimal point two places to the left. So in this example, you know that $\frac{40}{100}$ is the same as 40 with the decimal moved two places to the left, which means .4 is the decimal conversion.

Where did you end up? You ended up taking the original percent and dividing it by 100, or, in other words, moving the decimal point two places to the left. Knowing that, you can skip the fraction step in the middle and follow these steps: take the amount of the percent, drop the percent sign, and move the decimal point two places to the left.

Try a few more examples: Convert 14% to a decimal. Drop the percent sign and move the decimal point two spots to the left, which gives you .14. 50% turns into .50, which you know you can also write as .5.

Want to convert 8% to a decimal? Just move the decimal point two places to the left, which requires you to add a zero as a placeholder and gives you .08 as your answer. And if you're converting 115%, you'd have to move the decimal two places to the left, which would leave you with 1.15 as your answer.

FACT

If you want to move the decimal point more places to the right or left than the current digits allow, you can add zeroes to represent the empty places. For example, if you want to move the decimal point three places to the left on the number 9, you'd have to insert zeroes to give you .009. It may help to keep in mind that 9 has the same value as 000009.00000, you just never write those zeroes because they have no value.

Converting a Decimal to a Percent

Remember how, when converting a percent to a decimal, you move the decimal point two places to the left? If you do, it shouldn't surprise you to learn that to convert a decimal to a percent, you move the decimal point two places to the right.

Think about the number 1. The number 1 is 100% of 1. So the number .5 would be half of that, or 50% of the whole. And how do you get from .5 to 50? You move the decimal point two places to the right, which is the same as multiplying by 100.

You do that because when you convert something into a percent, you're converting it into the number it would be out of 100 pieces. When you have a decimal, you're displaying the information as if it's out of one whole, not 100 pieces. So if you want to change that decimal into a percent, you have to multiply it by 100.

Want to convert .8 to a percent? Just multiply it by 100; in other words, move that decimal point two places to the right. That gives you 80, so you know that .8 is the same as 80%. The common-sense check works: .8 is bigger than half but smaller than 1, and so is 80%. Want to convert .04 to a percent? Multiply it by 100, which gives you 4, so you know that .04 is equivalent to 4%.

Converting a Decimal to a Fraction

You can turn any number into a fraction without changing its value by putting the number in the numerator and sticking a 1 in the denominator.

For example, what if you want to represent .8 as a fraction? You would start by just throwing it in the numerator and sticking a 1 in the denominator. Now you have $\frac{.8}{1}$, which isn't exactly beautiful.

You want to get rid of the decimal in the numerator so that you have integers in the numerator and denominator. To change .8 into an integer, you could multiply it by 10—in other words, you could move the decimal one place to the right. Whichever way you like to think of it, you end up with 8 in the numerator. And because you are trying to alter the fraction without changing its value, you have to go ahead and multiply the denominator by 10 as well. Instead of 1, the denominator is now 10. So the .8 has turned into $\frac{8}{10}$. That's looking like a fraction! And now you can reduce that fraction by dividing out the common factor of 2, which simplifies the fraction to $\frac{4}{5}$.

Common Decimal–Fraction–Percent Conversions to Know

Sometimes, memorizing a few commonly used numbers can be really helpful. The chart below covers some common representations of decimals, percentages, and fractions. In each row the three entries are equivalent—or, in the case of two of the percents, approximately equivalent (≈). Even if you decide not to memorize this list, look it over and see how many of these you already know from your daily life.

Decimal	Fraction	Percent
.05	1/20	5%
.1	1/10	10%
.25	1/4	25%
.3333333 . . .	1/3	≈33.33%
.5	1/2	50%
.666666 . . .	2/3	≈66.66%
.75	3/4	75%
1	1	100%

Finding a Portion of an Integer

Sometimes you're asked to find a portion of an integer. You might be asked to find 30% of 20. You might be asked to find $\frac{1}{4}$ of 20. You aren't usually asked to find .2 of 20, because you don't really say, "What is point two of twenty?" Instead, you'd just be asked to multiply .2 by 20, and you already know how to do that.

Take a look at fractions first. When you're asked to find, for example, $\frac{1}{4}$ of 20, that means you should multiply $\frac{1}{4}$ by 20. So this is just a multiplication problem, like with any other fraction.

1. Turn the integer into a fraction by putting a 1 in the denominator. Now you have $\frac{1}{4} \times \frac{20}{1}$.
2. Multiply the numerators and the denominators: $\frac{1 \times 20}{4 \times 1} = \frac{20}{4}$.
3. Reduce the fraction: 20 and 4 both share a factor of 4, so $\frac{20}{4}$ reduces to $\frac{5}{1}$, which can be written more simply as 5.

Percent has one extra step. As you already know, you can convert a percent to a fraction by putting the percent number over 100. So 30% is the same as $\frac{30}{100}$. You can reduce that to $\frac{3}{10}$. Now that you have the percent in fraction form, you just use it like a fraction. Finding 30% of 20 is the same as finding $\frac{3}{10}$ of 20, and following the fraction steps gives you $\frac{60}{10}$, which reduces to 6.

Chapter 6 Exercises

Convert each fraction to a decimal. If necessary, round to two decimal places.

1. $\frac{3}{9}$
2. $\frac{4}{8}$
3. $\frac{2}{5}$
4. $\frac{6}{15}$
5. $\frac{3}{12}$
6. $\frac{40}{10}$

Convert each fraction to a percent. If necessary, round to the nearest percent.

1. $\frac{3}{4}$
2. $\frac{4}{5}$
3. $\frac{8}{6}$
4. $\frac{3}{2}$
5. $\frac{1}{5}$
6. $\frac{20}{35}$

Convert each percent to a decimal. Leave off any unnecessary zeroes.

1. 50%
2. 13%
3. 122%
4. 4%

Convert each decimal to a percent.

1. 4.5
2. .06
3. .23
4. .1

Convert each decimal to a fraction. Write your answer in the most simplified form.

1. .2
2. 1.5
3. .06
4. .24
5. .12
6. 1.05

Convert each percent to a fraction. Write your answer in the most simplified form.

1. 40%
2. 15%
3. 2%
4. 140%
5. 30%
6. .2%

CHAPTER 7

The Four Functions

The four functions are addition, subtraction, multiplication, and division. Since you've already made it to pre-algebra, you are probably pretty good at doing these things already. You do have to make sure that you are clear on the rules of executing these functions on all rational numbers. In this chapter, you'll learn how to express these rules in general terms so that they can be applied to variables later.

Adding Integers

When you add two positive integers, you just sum their value. No big news here. When you add zero to any integer, the value of that integer stays the same. So, for example, $4+0=4$ and $-6+0=-6$.

When you add two negative integers, you add together the absolute value of those integers and keep the negative sign. For example, $-4+-6=-10$. Because both numbers have a negative sign, they are both moving in the same direction on the number line. So this equation asks for the cumulative effect of negative four and negative six. That's negative ten. For computation's sake, when adding numbers with the same sign, simply add them and keep the sign. Just as $4+6=10$, it's also true that $-4+-6=-10$.

Adding Positive and Negative Numbers Together

When you add a positive number to a negative number, you aren't really doing addition. One of those numbers is moving you to the right on the number line, and one of them is moving you to the left. Thus, they work against one another, not together. To find their sum, you should ignore the sign and subtract the absolute values of the two numbers. Then, use the sign of the number with the bigger magnitude, or absolute value. For example, if you're asked to find the sum of –6 and 4, you should first notice that they have different signs. So instead of adding, you subtract their absolute values: $6-4=2$. Since –6 has a bigger magnitude (distance from zero) than 4, your final answer should start with a negative sign. Thus, the sum of $-6+4=-2$.

Subtracting Integers

When you subtract one positive integer from another, the result could be positive or negative. When faced with a subtraction problem, always subtract the smaller number from the bigger number to find their difference. If the problem has you subtracting a small number from a big number, the result will be positive. If the problem has you subtracting a big number from a small number, the result will be negative.

For example, if you're asked to find $6-4$, your solution will be the **difference** between these two numbers, which is 2. You're done! However, if you're asked to find $4-6$, you know the result will be negative, because a bigger number is being subtracted from a smaller number. You still find their difference by subtracting $6-4$, which is 2. Since you know the result will be negative, the difference is –2.

When you subtract zero from any number, there is no effect on the value. So $4-0=4$, for example, and $-6-0=-6$.

Examples of Subtracting Integers

When you subtract something from zero, its absolute value stays the same, but its sign changes. Thus, $0-6=-6$ and $0-(-6)=6$. Note that the second example is the same as $0+6$.

When you subtract a negative number from another negative number, it is easiest to change the problem into an addition problem. Subtracting a negative number is the same as adding a positive one. So for example, you would change $-4-(-6)$ into $-4+6$. Now you have an addition problem, and you can go back to your rules about adding numbers with different signs. You subtract their absolute values and keep the sign of the number with the bigger absolute value. In this example, 6 has a bigger absolute value than –4, so the sign of the answer must be positive. Thus, $-4-(-6)=2$.

When you subtract a negative number from a positive number, it is easiest to change the problem into an addition problem. For example, $4-(-6)$ turns into $4+6$. Now you have the simple addition of two positive numbers.

When you subtract a positive number from a negative number, you are basically adding two negative numbers. For example, when asked to find $-4-6$, you're starting with a negative number and moving even farther to the left on the number line. It's the same as adding –4 and –6, which you do by adding 4 and 6 and keeping the negative sign. Thus, $-4-6=-10$.

Adding and Subtracting with Non-Integers

All the previously mentioned rules of positive, negative, and zero work whether the numbers you're adding and subtracting are integers or not. When you have an addition or subtraction problem where some numbers are integers and others are fractions, you will usually find it easiest to add and subtract all the integers first and then add and subtract all the fractions, leaving you with one integer and one fraction that have to be combined. To combine them, you must convert the integer into a fraction with the same denominator as the fraction, so that your two fractions can be added or subtracted.

Here's an example. If you're asked to find $4+\frac{3}{2}-9-\frac{5}{6}+3$, start by rearranging the expression so that the integers are grouped and the fractions are grouped. You have to be sure to take your subtraction signs with the numbers that follow them so that you don't change the overall value of the expression. So your first step is to rewrite the expression as $4-9+3+\frac{3}{2}-\frac{5}{6}$. You can combine the integers to get –2. To combine the fractions, you need to give them a common denominator. Since the LCM of 2 and 6 is 6, you should convert both fractions to sixths. The fractions in the expression convert to $\frac{9}{6}-\frac{5}{6}$, which simplifies to $\frac{4}{6}$. You can reduce that fraction to $\frac{2}{3}$.

Now you have simplified the expression to $-2+\frac{2}{3}$. In order to combine these, you need to make the integer into a fraction by placing it over 1, and then you can give the fractions a common denominator. Once you do that, you have $-\frac{6}{3}+\frac{2}{3}$. Because you are adding two numbers with different signs, you subtract the numbers, $\frac{6}{3}-\frac{2}{3}=\frac{4}{3}$, and keep the sign of the number with the bigger magnitude, so the final value is $-\frac{4}{3}$.

If you're asked to do addition or subtraction with a mix of integers and decimals, the math is a little simpler. You don't have to convert anything—you just follow the integer rules of addition or subtraction, but with more place values.

Multiplying Integers

By now you're pretty comfortable multiplying integers. You know that when you multiply two positive integers, their product is positive. When you multiply two negative integers, their product is also positive. When you multiply a negative by a positive, or vice versa, their product is negative. When you multiply any number, negative or positive, by zero, their product is zero.

Symbols

When you first start learning math, the most common symbol used to show multiplication is ×. It's not that this symbol is wrong; it's just not used much once you get to pre-algebra. That's because it looks an awful lot like an *x*, which is the most common variable used in algebra. You can't have *x* and × getting confused in a problem.

What should you use instead? You have three other options for symbolically representing that you are looking for the **product** of two numbers.

1. **You can use a dot symbol to show multiplication.** Instead of writing 4×6, you can write $4 \cdot 6$. It means exactly the same thing, but you won't be worried about that dot getting confused with a variable. Make sure you don't put the dot so low that it looks like a decimal point! If it helps, you can write the dot a little bigger, like $4 \bullet 6$.
2. **You can use parentheses to show multiplication.** Instead of writing 4×6, you can write $4(6)$ or $(4)(6)$. When you put two things in parentheses next to one another, it tells you that those numbers are being multiplied. You have the option of putting parentheses around only the second term being multiplied if the first term being multiplied is a single number. For example, you can write $5(9+3)$ or you can write $(5)(9+3)$, and both are the same as writing $5 \times (9+3)$.
3. **You can sometimes put two terms next to one another to show multiplication.** You can't do this all the time, because sometimes it would be confusing. Obviously if you want to write 5×4, you can't write it as 54. How would you know that it wasn't just the number 54? You wouldn't.

   But when the things you're multiplying aren't numbers, you can just put them next to one another. You aren't going to see variables (letters that stand for numbers) for a few more chapters, but there are some numerical examples of this method. For example, instead of writing $5^4 \times 3^2$, you can leave out the symbol and just write $5^4 3^2$. There's no confusion about which numbers are being multiplied, because the exponents break up the expression. This is one example of a time when you can leave out the multiplication symbol and just put two things next to one another as an indication that they are supposed to be multiplied together.

Dividing Integers

When you divide one integer by another, the first one is called the **dividend** and the one you're dividing by is called the **divisor**. So if you divide 4 by 2, then 4 is the dividend and 2 is the divisor. The solution to a division problem is called the **quotient**. Up until now, you've mostly seen the symbol for division as $\div$. So you'd be asked to find $8 \div 2$ in order to get the quotient of 4.

Symbols

The $\div$ symbol isn't wrong, but you won't really use it much in pre-algebra and beyond. You're entering the world of algebra now, and so you're getting to a more advanced level of the math "code." You could still write all your division with the $\div$ symbol, but you shouldn't, because there is an alternate way of writing division that makes your life a *lot* easier when you are doing algebra.

Instead of using that division symbol, you should switch to using a fraction bar. You've actually been using this symbol for division ever since you learned about fractions. When you write $\frac{1}{2}$, you're really saying to take one thing and divide it into two pieces. What do you end up with? Half a thing. That's why 1 divided by 2 gives you the decimal of .5, which has the same value as $\frac{1}{2}$.

Fractions are basically division. Think about the fraction $\frac{8}{2}$. You know this fraction isn't in its most reduced form. The numerator and denominator have a common factor of 2, which you can factor out, simplifying the fraction to $\frac{4}{1}$. And you know that $\frac{4}{1}$ is the same as 4, because a 1 in the denominator has no effect on the size of the fraction. Why? Because it's the same as dividing something by 1, which has no effect. So the fraction $\frac{8}{2}$ simplifies to 4. This proves the idea that the fraction bar is just telling you to divide the numerator by the denominator: $8 \div 2 = 4$.

The fraction bar is going to work as your new division symbol. The number above the bar gets divided by the number below the bar. If the top number is bigger than the bottom number, your quotient will be greater than 1. For example, $\frac{10}{5} = 2$. If the number on the top is smaller than the bottom number, your quotient will be less than 1. For example, $\frac{10}{40}$ is equal to $\frac{1}{4}$ or .25.

All the basic rules of division still apply even though the symbols have changed. When you divide a positive number by a positive number, the quotient is positive. When you divide a negative number by a negative number, the quotient is also positive. When either your dividend (the number on top) or your divisor (the number on the bottom) is negative and the other one is positive, your quotient will be negative.

Special Rules

As usual, there are a few special rules for 0 and 1. Dividing a number by 1 has no effect on the value of that number. When you divide something into one group, that group will be the same size as what you started with. When you divide zero by anything, the result is zero. You can divide zero evenly into as many groups as you want, as each group will always have exactly zero things in it. That's why $\frac{0}{8}$ is the same as $\frac{0}{10}$ or $\frac{0}{40}$ or $\frac{0}{-10}$. They're all equal to zero.

What about when you take a number and divide it by zero?
You can't. You'd have to make things disappear! It doesn't make sense in real life, and it doesn't make sense in math. That's why you can never have 0 as the denominator of a fraction.

Multiplying and Dividing with Non-Integers

While memorizing the rules of multiplying and dividing with non-integers can be a good place to start, you have to practice this stuff until it's just as easy for you as multiplying or dividing integers.

To make things easier, here are the rules for multiplying and dividing non-integers (decimals, fractions, and percent) by integers, one step at a time.

Multiplying Fractions by Integers

Turn the integers into fractions by placing a 1 in the denominator. Multiply the numerators to get your new numerator. Multiply the denominators to get your new denominator. Simplify the result by factoring out any factors shared by the numerator and denominator.

Dividing Fractions by Integers (or Vice Versa)

Turn the integers into fractions by placing a 1 in the denominator. Take the reciprocal of the divisor by flipping it over (switching the numerator and the denominator). Now, multiply the numerators to get your new numerator. Multiply the denominators to get your new denominator. Simplify the results by factoring out any factors shared by the numerator and denominator.

Multiplying or Dividing with Decimals and Integers

If you have a calculator, just plug in the numbers with their decimal points. If you are working on paper, multiply or divide as if the decimal places aren't there at all. Count up the digits to the right of the decimal point in any numbers you are *not* dividing by. Place the decimal point in your answer so that there are the same amount of digits to the right of the decimal point as there were, in total, in all the numbers you counted. Then, if you divided by any numbers with a decimal, count the digits to the right of the decimal point in all the numbers of the divisor. Move the decimal point that many places to the right in your answer. For example, say you're asked to multiply $1.4 \times .02$. Start by pretending the decimal places aren't there, and just multiply $14 \times 2 = 28$. Now, count the total digits to the right of the decimal places in all the numbers you multiplied (do not include any nonessential zeroes). 1.4 has one digit to the right of the decimal place, and .02 has two, which makes three decimal places in total. That means I have to move the decimal point in my answer three places to the left, turning 28 into .028, which is the answer.

Multiplying Integers and Percent

You don't really divide by percent; instead, you usually take a percent *of* a number, and *of* stands for multiplication. To multiply integers and percent, take any percent value and make it into a fraction by putting it over the number 100. Reduce the fraction if possible. Then, follow all the rules of multiplying fractions.

Chapter 7 Exercises

Answer the following questions with one word or a short phrase.

1. When you add two negative numbers, is the sum positive or negative?
2. When you add two positive numbers, is the sum positive or negative?
3. Is it possible to add a negative number and a positive number and get a negative sum?
4. Is it possible to add a negative number and a positive number and get a positive sum?
5. When you multiply two positive numbers, is the product positive or negative?
6. When you multiply two negative numbers, is the product positive or negative?
7. When you multiply a positive number by a negative number, is the product positive or negative?
8. What do you get when you multiply a number by zero?
9. What do you get when you divide a number by zero?
10. What do you get when you multiply a number by one?
11. What do you get when you divide a number by one?
12. When you subtract one integer from another, will your answer always be an integer?
13. When you add one integer to another, will your answer always be an integer?
14. When you multiply one integer by another, will your answer always be an integer?
15. When you divide one integer by another, will your answer always be an integer?

CHAPTER 8

Exponential Powers

Like most of the math in this book, the rules for working with exponents make sense when you understand them, because they're based on what happens with numbers in the real world. What's *hard* about exponents is getting used to what the symbols mean, what they're telling you to do, and what you are and are not allowed to do. While it's a good idea to memorize all these rules, memorization really isn't enough to make it stick. You also have to understand *why* the rules exist. Understanding where the rules come from makes you much more likely to remember them and apply them correctly.

Exponents

The most important step when it comes to working with exponents is to learn what they mean. Take a look at the expression 2^5. The 2 is called the **base**. The base is the big number at the bottom of an exponent. The 5 is called the **exponent**. The exponent is the little number above and to the right of the base.

You can talk about exponents in a few different ways. Take another look at 2^5. You can call that "two raised to the fifth power," or "two to the fifth" for short. If the exponent is a 2 or a 3, there are special words you can use. For example, you can call 5^2 "five raised to the second power" or "five to the second," but you can also call it "five squared." When you raise something to the second power, that's called "squaring" it. You can call 5^3 "five raised to the third power" or "five to the third," but you can also call it "five cubed." When you raise something to the third power, that's called "cubing" it.

What Do Exponents Mean?

The exponent tells you how many times the base will be used as a factor. For example:

2^1 tells you to take the base of 2 as a factor one time. Thus, $2^1 = 2$.
2^2 tells you to take the base of 2 as a factor two times. Thus, $2^2 = 2 \times 2 = 4$.
2^3 tells you to take the base of 2 as a factor three times. Thus, $2^3 = 2 \times 2 \times 2 = 8$.

2^3 is *not* the same as 2×3. 2^3 is not telling you to multiply two by three. It's telling you to take the number 2 and multiply three of that number together. In other words, the base is the factor being multiplied, and the exponent just tells you how many times that factor is used in computing the final value.

When you see an exponent, remind yourself that the big number on the bottom is the one you are multiplying—the little number up top just tells you how many times to multiply it.

Exponents are funny creatures. They cause numbers to get very big, very fast. This effect is called "exponential growth," which might be an expression you've heard before. Let's say that your friend makes you an offer. You can choose one of the following options: The first option is that he'll give you $2 today, and then every day for the rest of the year he'll give you $2 more than he gave you the day before. In other words, today you'll get 2×1 dollars; tomorrow you'll get 2×2 dollars; the next day you'll get 2×3 dollars; and so on for the rest of the year. Your second option is that today, he'll give you $2; tomorrow he'll give you twice what he gave you today; and then in two days, he'll give you twice what he gave you the day before. In other words, every day he will give you double what he gave you the day before. So today he'll give you 2^1 dollars; tomorrow he'll give you 2^2 dollars; the next day he'll give you 2^3 dollars; and so on. Which deal is the better option?

You can probably see that the second option is going to get you more money. Even though both options involve multiplying by 2, in the first option you are multiplying 2 by however many days have gone by, whereas in the second option, you're multiplying 2 by whatever you had the day before. The results start to compound . . . a lot. By the end of the year, option one is going to net you a tidy profit of $132,925. Seems like a lot. Option two, however, is going to make you really rich: your profit would be $150,306,725,297,253,300,000,000,000,000,000,000,000,000,000,000,000,000,000,000,000,000,000,000,000,000,000,000,000,000,000,000,000,000,000,000,000,000. When you grow numbers in an exponential way, they get bigger a lot faster than when you grow them through multiplying by other numbers. For example: $2 \times 10 = 20$, but $2^{10} = 1024$. Why does it get so big so fast? Because 2×10 is just ten groups of 2. In other words, it's 2 being added to itself 10 times. (If you really want, you could do out $2+2+2+2+2+2+2+2+2+2$ and see that it's the same.) But 2^{10} is a factor of 2 multiplied ten times, which means its value is equal to $2 \times 2 \times 2 \times 2 \times 2 \times 2 \times 2 \times 2 \times 2 \times 2$.

Special Rules to Remember

Here are a few things to remember about exponents. Keep in mind that none of these notes change the rules at all—they are just special things to know and remember because they show up a lot during pre-algebra.

1. Whenever you have an exponent of 1, that's the same as having no exponent at all. The base just stays as whatever it was, and that's it. $10^1 = 10$, $15^1 = 15$, $(-1)^1 = -1$. No matter what the base is, if you only have 1 of it, the value doesn't change at all.

Sometimes you'll see the 1 written as an exponent, and sometimes you won't. The value is the same, but usually if a number simplifies to have 1 as its exponent, you should just write down the number. However, sometimes you'll need to include the exponent to clarify a point or when you're specifically asked to.

2. Whenever you raise something to an exponent of 0, the result is equal to 1. You can think of it this way: when you multiply something by 2^1, you're just multiplying it by the number 2 once, which is the same as multiplying it by 2. But when you multiply something by 2^0, you're multiplying it by the number 2 zero times, which has absolutely no effect. It's the same as multiplying something by 1, which is why anything to the zero power is equal to 1.

Watch out for zero as an exponent, and don't let it trick you. No matter what the base is, zero as an exponent turns the result into 1.

3. When the base is 1, it doesn't matter what the exponent is: the result will still be 1. That's because you can multiply 1 by itself as many times as you want, and the value doesn't change.
4. When the base is a fraction, the rules don't change! Just take the fraction, and use it as a factor, however many times the exponent tells you to do so. For example, $\left(\frac{1}{2}\right)^3$ is the same as $\frac{1}{2}\times\frac{1}{2}\times\frac{1}{2}$. As always, follow your

fraction rules, multiplying the numerators to get your new numerator and multiplying the denominators to get your new denominator, which gives you $\frac{1}{8}$.

5. When the base is positive, the result will always be positive. That's because when you multiply all positive numbers together, it doesn't matter how many you have; the result will be positive.
6. When the base is negative, watch out! If you raise a negative base to an even power, the result will be positive. Why? Because when you multiply two negatives together, the product is positive. And if your exponent is even, every negative number has a matching negative number to be multiplied by, leaving you with all positive numbers. However, if the base is negative and the exponent is odd, the result will be negative. Why? Take a look at $(-2)^3$ as an example. This turns into $(-2)(-2)(-2)$. Two of those negatives cancel one another out and give you a positive: $(-2)(-2)=4$. Once you multiply by that final –2, you're multiplying a positive 4 by a negative 2, which leaves you with a product of –8.

FACT

When you have a negative base, look at the exponent to see if it is even or odd. If the exponent is even, your final result will be positive. If the exponent is odd, your final result will be negative.

Negative Exponents

Negative exponents are a little tricky. They don't actually have anything to do with negative numbers. Remember, math is a code, and exponents are only a tricky part of that code because they're newer to you than the rest of the code. You're used to seeing numbers that are next to one another being multiplied, so it's easier to think of multiplication. You're used to the negative sign next to a number meaning that the number is negative, so that seems to be its logical meaning. But a code can mean whatever you want it to mean, as long as everyone agrees on the rule. Here's the rule for negative exponents.

Look at the example of 2^3. That means you have the number 2 as a factor three times, so you can think of 2^3 as $2\times2\times2$, which equals 8. What about 2^{-3}? Now the exponent is negative, which changes things. A negative exponent tells you to take the base and *factor it out*, instead of listing it as a factor, however many times the exponent indicates. Thus, just as 2^3 is the same as $2\times2\times2$, 2^{-3} is more like $\frac{1}{2}\times\frac{1}{2}\times\frac{1}{2}$.

Dividing something by 2 three different times is the same as dividing it by 8, so multiplying by 2^{-3} is the same as dividing by 8, which is the same as multiplying by $\frac{1}{8}$. Because of this, $2^{-3}=\frac{1}{8}$. Since "divide by 2" is the same as "multiply by $\frac{1}{2}$," it's probably easier to think about negative exponents in this way. When you see 2^{-3}, you can think of it as $\left(\frac{1}{2}\right)^3$. Notice what happened here: you dropped the negative sign off the exponent and flipped the base over. That's because the negative sign on the exponent told you to think in terms of division, so you've just traded that negative sign code for division code. Now that you have $\left(\frac{1}{2}\right)^3$, you can follow all the normal rules of exponents—take the base and multiply it however many times the exponent tells you to do so. So $\left(\frac{1}{2}\right)^3$ is the same as $\frac{1}{2}\times\frac{1}{2}\times\frac{1}{2}$, which is equal to $\frac{1}{8}$.

Here's a review of the steps for dealing with negative exponents: Drop the negative sign off the exponent. Replace the base with its reciprocal (meaning, "flip" the fraction over). Complete the exponent problem as usual.

Perfect Squares

A **perfect square** is a number that can be made by squaring an integer. In other words, it's a number that you can get by raising an integer to the second power, or by taking an integer and multiplying it by itself. It can be really helpful to learn the perfect squares by memorizing them. Knowing that a number is a perfect square can make your life easier in pre-algebra, algebra, and beyond. It can really be a big help when it comes to estimation.

Perfect Square Examples

The smallest perfect square is 0. When you take zero and square it, you end up with 0^2, which is the same as 0×0, which equals 0. It is probably worth your time to memorize the first 15 perfect squares, just so that you are comfortable with them.

$0 \times 0 = 0$
$1 \times 1 = 1$
$2 \times 2 = 4$
$3 \times 3 = 9$
$4 \times 4 = 16$
$5 \times 5 = 25$
$6 \times 6 = 36$
$7 \times 7 = 49$
$8 \times 8 = 64$
$9 \times 9 = 81$
$10 \times 10 = 100$
$11 \times 11 = 121$
$12 \times 12 = 144$
$13 \times 13 = 169$
$14 \times 14 = 196$
$15 \times 15 = 225$

Roots

Math is full of concepts that work together because they are opposite sides of the same idea. For example, addition and subtraction share a lot of rules and properties because they are very similar. In fact, subtracting is just adding negative numbers, which means it is just another form of addition.

Multiplication and division share a lot of rules and properties as well. Why? Because dividing is really just multiplying by a reciprocal, which means they are really both forms of multiplication. Exponents and roots are the same way: they have a lot of properties and rules in common, because roots are really just another form of exponents.

Finding the Root

The most common root, and maybe the only one you'll deal with in pre-algebra, is a square root. The symbol for a square root looks like this: $\sqrt{\ }$. That symbol is sometimes called a **radical**. It doesn't show up in math just floating like that, though; it has to have something underneath it. So it might look, for example, like this: $\sqrt{4}$. You would read that expression as "the square root of four" or sometimes even "root four" for short.

When you put something under the square root symbol, you're basically asking the question: "What number, when multiplied by itself, would give you this value?" In other words, the expression $\sqrt{4}$ stands for whatever number yields 4 when multiplied by itself. Therefore, a root is basically the opposite of an exponent.

When you "square" 4, or simplify the expression 4^2, you're figuring out the question "What number do you get when you multiply 4 by itself?" When you "root" 4, or simplify the expression $\sqrt{4}$, you're answering the question "What number can you multiply by itself to give you 4 as the answer?" The answer, of course, is 2. When you multiply 2 by itself, you get 4 as the result. For that reason, $\sqrt{4} = 2$.

See why knowing those perfect squares is helpful? Once you recognize a perfect square under the radical, you already know what the root is. You know that $\sqrt{100} = 10$ because you already memorized that $10 \times 10 = 100$.

Notice that, most of the time, when you take the square root of a number, it gets smaller. When you try to simplify $\sqrt{64}$, for example, you're trying to find what number, when multiplied by itself, will give you 64 as the product. In other words, $\sqrt{64}$ is equal to whatever number, when multiplied by itself, yields a product of 64. That number has to be smaller than 64. And, as you know from our memorization list, $\sqrt{64} = 8$. If the number under the radical is bigger than 1, its square root will be smaller than it is. That's because if you want to multiply something by itself to get a number bigger than 1, the thing you're multiplying has to be smaller than the value you end up with.

$\sqrt{1}$ is a little different. This expression is equal to the number that, when multiplied by itself, yields a product of 1. And what number does that? The number 1 does. Because $1 \times 1 = 1$, $\sqrt{1} = 1$. So 1 is a very interesting number, because 1 = $\sqrt{1}$. The number 1 doesn't change no matter how many times you multiply it by itself or divide it by itself. That's why

$1 = 1^2 = 1^5 = 1^{100} = 1^{-10} = 1^{-40}$. It doesn't matter what the exponent is at all: if the base is 1, the expression is equal to 1.

More than Just Square Roots

Remember that when you square something, that means you're raising it to the power of 2. When you take the square root of a number, you're trying to figure out what number, when raised to the power of 2, gives you the value you started with. But 2 isn't your only exponent. You can "cube" something, raising it to the third power, or you can raise it to the fourth power or the tenth power or the twentieth power, depending on whether the exponent is 4 or 10 or 20.

The same is true for roots. Another way to write the $\sqrt{\ }$ symbol is to write $\sqrt[2]{\ }$. Those symbols mean exactly the same thing; they both mean "square root." You will sometimes leave out the little 2 because the square root is by far the most common root to find. It's sort of the default choice, so you don't have to bother clarifying. It's the same reason you'll sometimes hear "root 4" instead of "square root of 4." Unless you specifically bother to say otherwise, the root you're talking about is the square root.

If this seems weird to you, think about the negative sign. You don't bother to put a positive sign in front of all the positive numbers. Rather, in the absence of a negative sign, the positive sign is assumed. The idea for this shortcut is to save you time and energy, although sometimes these shortcuts can make it confusing to learn new symbols and what they mean.

Just as $\sqrt[2]{16}$ simplifies to 4 (the number that, when multiplied by itself, yields a product of 16), so $\sqrt[4]{16}$ is 2, because $2 \times 2 \times 2 \times 2 = 16$. If you wanted to talk about that symbol, you'd call it the "fourth root." You won't usually see many of these, but you will occasionally be asked to simplify the "third root" or "cube root" of something. For example, how could you simplify $\sqrt[3]{8}$? You need to know what number, when multiplied together 3 times, will give you 8. The answer is 2, because $2 \times 2 \times 2 = 8$. Thus $\sqrt[3]{8}$, or "the cube root of 8," is 2.

Writing a Root as an Exponent

Remember how, at the beginning of this section, it was stated that roots are really just a form of exponents? They are. You can already see how they are related, but you can actually write a root as an exponent anytime you want. You might not see this in class a lot, but it's helpful to know, because it helps drive home the point that the same rules that apply to exponents apply to roots: the same rules have to apply, because roots are really just a special type of exponent.

Let's say that you want to write "the square root of 100" as an expression. You can write $\sqrt{100}$, but you can also write $100^{\frac{1}{2}}$ if you want to. That's because raising something to the "one half" power is the same as taking its square root. Why? Well, when you raise a number to the second power, you're saying, "What happens when you multiply this number by itself?" So when you raise a number to the one-half power, you're saying, "What do you have to multiply by itself to get this number?" Fractional exponents don't really have much to do with fractions—it's not like they turn the base into a fraction or give you a fraction as the result. Fractions stand for a different code when they are up in the exponent: they stand for roots. So if you wanted to, you could write $\sqrt[3]{8}$ as $8^{\frac{1}{3}}$, because these two sets of symbols mean the exact same thing.

Estimating Roots

You might be asked to estimate square roots as part of a quiz or test. Estimating roots is a useful skill to have as you continue your study of algebra, although it may seem silly now. Remember that the trick with estimating is always to think about something you know that's close to what you're being asked to work with. For example, if you're asked to estimate $99 \div 5$, you're not going to divide out $99 \div 5$ and then round your result. Instead, you're going to think about $100 \div 5$, which you know by memory is equal to 20.

How does this apply to roots? Well, what do you do when you're asked to estimate $\sqrt{99}$? You don't know the square root of 99, and you don't even have a way to figure it out using arithmetic. That's why estimating roots can be so important—unlike your sample division problem, there's really no way you can algebraically solve $\sqrt{99}$ without a calculator. You can simplify it, but you can't just turn it into a number. You have to think about a perfect square that's close to 99 and base your estimate on that. Since you know $\sqrt{100} = 10$, you know that $\sqrt{99}$ has to be pretty darn close to 10. Actually, you know it has to be a little bit smaller than 10, because 99 is a little bit smaller than 100. In fact, the square root of 99 is approximately 9.95, which really is pretty close to what you guessed!

What if you're asked to estimate $\sqrt{6}$? Well, what perfect squares do you know that are close to 6? You know that 4 is a perfect square (it's 2^2), and you know that 9 is a perfect square (it's 3^2), and 6 is somewhere between 4 and 9. Therefore, the square root of 6 must be somewhere between the square root of 4 and the square root of 9. In other words, the square root of 6 must be somewhere between 2 and 3. Which one is it closer to? Well, 6 is closer to 4 than it is to 9, because 6 and 4 have a difference of 2, whereas 6 and 9 have a difference of 3. This means that $\sqrt{6}$ must not only be between 2 and 3; it must be closer to 2 than 3. You might estimate that $\sqrt{6}$ is 2.4 or somewhere around there. And you'd be right: $\sqrt{6}$ is approximately equal to 2.45.

If you're asked to estimate the square root of a number, you should think about where it fits into the list of squares mentioned earlier in this chapter. For example, if you're asked to estimate $\sqrt{200}$, you're going to imagine the list and notice that 200 is between 196 and 225, so the square root of 200 is between 14 and 15. Since 200 is closer to 196 than it is to 225, $\sqrt{200}$ is closer to 14 than 15. Therefore, you might estimate it to be something like 14.3.

To estimate the root of a number, find the two perfect squares that number falls between. Then, take the square root of each perfect square. The root of your number will be somewhere between the square roots of the perfect squares it falls between.

Radical Numbers

Remember that another word for *root* is **radical**. So, when you see $\sqrt{2}$, you can call it "square root of two" or "root two" or "radical two." You already know how to simplify a radical if it's the radical of a perfect square. You also already know how to estimate a radical based on its proximity to perfect squares. But what you don't know yet is how to simplify a radical number.

Some radical numbers can't be simplified. Radical 2, or $\sqrt{2}$, is a good example. You can estimate $\sqrt{2}$ if you want—it is approximately equal to 1.4—but you can't actually represent it as a fraction or a decimal. $\sqrt{2}$ is 1.4142135623731 . . . it just goes on forever! It's an irrational number, so you'll usually leave it in the $\sqrt{2}$ form unless you're estimating. In short, $\sqrt{2}$ is the simplest way to write and express that value without estimating, so there's not much you can do with it.

Generally, when you simplify numbers, you will want to write them in the way that's easiest to understand without changing the value of the number. You will often leave $\sqrt{3}$ or $\sqrt{5}$ as they are, because there's no useful way to simplify radical 3 or radical 5. Keep in mind, though, that this doesn't mean there's no way to simplify a radical just because the number under the radical isn't a perfect square.

Chapter 8 Exercises

Estimate the following square roots to one decimal place. Do not use a calculator.

1. $\sqrt{3}$
2. $\sqrt{50}$
3. $\sqrt{99}$
4. $\sqrt{30}$
5. $\sqrt{200}$
6. $\sqrt{20}$

Rewrite the following expressions so that each expression is an integer.

1. 2^3
2. 9^1
3. 1^{14}
4. 6^0
5. $\sqrt{81}$
6. $\sqrt{1}$
7. 2^2+2^2
8. 5^2-5^2
9. $5^2\times3$
10. $4+4^2$
11. $1-1^5$
12. 9×8^0
13. 9×0^4
14. $6+\sqrt{100}$
15. $\sqrt{64}+\sqrt{64}$
16. $2\times\sqrt{81}$

CHAPTER 9

Manipulating Exponential Expressions

Now that you know what exponents are and how they work, you can learn how to "do math" with them. Most of the math you'll do involves manipulating an expression: adding or multiplying it, for example. This chapter will cover how you can combine and rewrite exponents and root expressions in different ways.

Multiplying with Exponents

It will be easiest to start with an example where the bases are the same but the exponents are different. What if you're asked to simplify $5^4 \times 5^6$? If you had a calculator, you could find both values, multiply them, and write down the product. But you'll be dealing with some *big* numbers. You're probably not going to be expected to find 5^6 by hand. So when you're asked to simplify $5^4 \times 5^6$, you are just being asked to write it in a simpler format. Let's look at what $5^4 \times 5^6$ really means.

$$5^4 = 5 \times 5 \times 5 \times 5$$
$$5^6 = 5 \times 5 \times 5 \times 5 \times 5 \times 5$$

So you've got four 5s being multiplied by six 5s. In other words, you have ten 5s being multiplied together. You can think of it like this:

$$5^4 \times 5^6 = (5 \times 5 \times 5 \times 5) \times (5 \times 5 \times 5 \times 5 \times 5 \times 5)$$

Because all the fives are just being multiplied together, the parentheses don't mean anything, and you could rewrite the expression as: $5 \times 5 \times 5 \times 5 \times 5 \times 5 \times 5 \times 5 \times 5 \times 5$. What's a simpler way to write that? Well, it's the number 5 multiplied as a factor ten times. Fortunately, you know the code for that now: 5^{10}.

See what happened? It turns out that $5^4 \times 5^6 = 5^{10}$. You just added the exponents together. Why? Because the exponents count how many times something is multiplied as a factor. And if you multiply it as a factor four times and then multiply it as a factor another six times, you're really just multiplying it as a factor ten times in total. The rule to remember from this is that when numbers with the same base are being multiplied, you can add the exponents to simplify.

For one last example, say you're asked to simplify $4^8 \times 4^3 \times 4^6$. The bases are the same, so you simply keep the base as it is and add the exponents: $4^8 \times 4^3 \times 4^6 = 4^{(8+3+6)} = 4^{17}$.

Different Bases

What do you do if the bases aren't the same? Here is an example where the bases aren't the same, but the exponents are: $8^3 \times 2^3$. First, write out $8^3 \times 2^3$ without exponents to see what it really means in a code you're more comfortable with: $8^3 \times 2^3 = 8\times8\times8\times2\times2\times2$. You can put these numbers in whatever order you want. So you could rewrite the expression as $8^3 \times 2^3 = (8\times2)(8\times2)(8\times2)$.

Now you have (8×2) being multiplied as a factor three times, which you could rewrite as $(8\times2)^3$. Therefore, you know that $8^3 \times 2^3 = (8\times2)^3$, which you can further simplify to 16^3. You can probably see now that the shortcut for multiplication with exponents, if the bases are different but the exponents are the same, is to multiply the bases first and then raise that product to the original exponent.

FACT

When multiplying numbers with exponents, there are a few good shortcuts you can take. If the bases are the same, keep the base the same and add the exponents. If the bases are different but the exponents are the same, multiply the bases and keep the exponent.

What happens when the exponents are different and the bases are also different? Well, that makes simplifying sort of hard to do. There's no nice shortcut. The only time you'd be asked to do something like this is if the exponents in the expression were numbers you could figure out yourself. For example, let's say you're asked to simplify $2^3 \times 1^8$. The exponents are different, and the bases are different, so you don't have any tricks you can use. But that's okay! Because you don't need tricks here—you can just follow the order of operations here and simplify each exponential term first. You can think of 2^3 as $2\times2\times2$, which is 8. So now you can rewrite the expression as 8×1^8. Okay, what about 1^8? Well, if you don't remember the trick about 1 and exponents, you could rewrite this expression as $1\times1\times1\times1\times1\times1\times1\times1$. It's pretty easy to recognize that this expression has a value of 1. It's even easier to remember that 1 raised to any exponent will still be 1. Knowing what you know now, you can rewrite the expression as 8×1, which you know is equal to 8.

Multiplying with Square Roots

Multiplying with square roots seems like it would be harder than multiplying with exponents, but it's actually a little bit easier. With exponents, you have to worry about whether or not the exponents are the same before you use the shortcut. With square roots, you don't have to take that step. Remember, placing the symbol $\sqrt{\ }$ over a number is the same as taking that number and raising it to the $\frac{1}{2}$ power. In other words, $\sqrt{16} = 16^{\frac{1}{2}}$ and $\sqrt{9} = 9^{\frac{1}{2}}$. So whenever you have two square roots, you can follow the rule for multiplying when the exponents are the same . . . because they are!

What is the rule for multiplying when the exponents are the same? You multiply the bases and keep the exponent just like it is. So now you can do the same thing with square roots. When you are multiplying with square roots, you should multiply the numbers underneath the square root signs, and then put that product under the same square root sign. For example, let's say that you're asked to simplify $\sqrt{4} \times \sqrt{25}$. If you wanted to, you could solve each step first, because you recognize these perfect squares.

Because $\sqrt{4} = 2$ and $\sqrt{25} = 5$, the original expression could be rewritten as 2×5, which is equal to 10. This is a good way to double-check, and prove your shortcut. The shortcut says that when you're multiplying with square roots, you can multiply the numbers underneath the square roots and keep the square root symbol the same. In other words, it says that $\sqrt{4} \times \sqrt{25} = \sqrt{(4 \times 25)}$. Go ahead and follow the order of operations to see if that's correct.

You start with the parentheses, and you're able to simplify this expression to $\sqrt{100}$. Next in your order of operations are exponents and roots, and you know from the list of perfect squares that $\sqrt{100} = 10$. So the shortcut gave you the same answer as the original method, which proves that it works!

If you're multiplying with roots, check to make sure they are both square roots first. You can't use this shortcut if one is a square root and one is a cube root, for example. As long as they are both square roots, you can multiply what's under the roots and take the square root of that value.

You probably recognized $\sqrt{4}$ and $\sqrt{25}$ because you know their value. But what if you were asked to simplify $\sqrt{2} \times \sqrt{50}$? Well, you know that $\sqrt{2}$

is an irrational number that can't be written as an integer. And you don't know $\sqrt{50}$, although you do know it represents a value little bigger than 7 (because of your estimations, you know that $\sqrt{49}=7$). So you could estimate this answer, but you'll probably have an impossible time doing so. On the other hand, if you use your shortcut, you can rewrite $\sqrt{2}\times\sqrt{50}$ as $\sqrt{(2\times 50)}$. Now you can follow the order of operations and further simplify to get $\sqrt{100}$, and therefore you know that $\sqrt{2}\times\sqrt{50}=10$.

Dividing with Exponents and Roots

The rules for dividing with exponents and roots are a lot like the rules for multiplying with exponents and roots. As always, if you forget the rules, you can work out what the rule must be, as long as you understand what the symbols of the math code mean.

For example, start by trying to simplify $8^9 \div 8^7$. What's the rule? Well, try to figure it out by understanding what this expression really means. You are multiplying the number 8 as a factor nine times, and then you're dividing it by itself seven times. To make it easier to visualize, trade that division sign for a fraction bar and rewrite this expression as $\frac{8^9}{8^7}$. Now, just to make sense of what's going on here, you're going to write out each of those symbols as what it really means. So instead of writing it as $\frac{8^9}{8^7}$, you're going to write it as $\frac{(8\times 8\times 8\times 8\times 8\times 8\times 8\times 8\times 8)}{(8\times 8\times 8\times 8\times 8\times 8\times 8)}$.

If you ever get confused about an exponent problem, try writing out what it means to help you get unstuck!

Now you just have a fraction, and you can follow the fraction simplification rule you always follow and factor out any factors that are shared by both the numerator and the denominator. In this case, both the numerator and the denominator have seven 8s in them. When you factor those out, you are left with $\frac{8\times8}{1}$, which you can just rewrite as 8×8. If you wanted, you could simplify further and write your answer as 64. Or, if you were asked to write your answer in the form of an exponential expression, you could change 8×8 to 8^2. In the end, $8^9 \div 8^7 = 8^2$. Why? Because seven of the 8s canceled each another out, and only two of them remained.

It's good to go through that process so that you understand where the rules come from. But you won't be able to do that all the time. You need a rule that you can remember, and that rule is as follows: When dividing exponential expressions with the same base, keep the base the same and subtract the exponents. When you did the previous problem, you found that $8^9 \div 8^7 = 8^2$. See what happened? When you divided these exponential expressions with the same base, the base stayed the same, and you subtracted the exponents.

Notice that this is the reverse of the rule that says when you *multiply* exponential expressions with the same base, you *add* the exponents. Why? Because when you divide, you're making a fraction and cancelling out a lot of shared factors, and the subtraction shows you how many of the base will be left when you're finished.

Dividing Roots

Dividing with square roots is a lot like multiplying with square roots. Remember how you showed that $\sqrt{2}\times\sqrt{50}=10$? When you multiply two square roots, you can rewrite the problem by putting the multiplication problem under one square root sign. For example, $\sqrt{2}\times\sqrt{50}=\sqrt{(2\times50)}$. The good news is that you can do the same thing with division! So, for example, let's say that you're asked to simplify $\sqrt{8}\div\sqrt{2}$. You might confront the same problem in the format of $\frac{\sqrt{8}}{\sqrt{2}}$. You can rewrite this problem by moving the division under the radical and changing it to $\sqrt{\frac{8}{2}}$. That's good, because you know that $\frac{8}{2}=4$, so you know that $\sqrt{\frac{8}{2}}=\sqrt{4}$ and that $\sqrt{4}=2$. You're finished! $\sqrt{8}\div\sqrt{2}=2$.

Want some proof? What if you were asked to simplify $\sqrt{25}\div\sqrt{25}$? Easy, right? You already know that $\sqrt{25}=5$, so you can rewrite the expression $\sqrt{25}\div\sqrt{25}$ as $5\div5$, which is equal to 1. If your new shortcut is right, it should spit out the right answer when you follow the steps. Let's check. The shortcut says that you can rewrite $\sqrt{25}\div\sqrt{25}$ as $\sqrt{\frac{25}{25}}.\ \frac{25}{25}=1$, so you can rewrite this expression as $\sqrt{1}$, which simplifies to 1. That's good proof that the shortcut works!

Simplifying Radicals

Now that you know how to multiply and divide with square roots, you're ready to learn how to simplify radicals. Remember, a radical is just a number that's expressed by a value written under a root sign. You already know how to simplify some radicals. For example, you know that you can rewrite $\sqrt{100}$ as 10 and you can rewrite $\sqrt{25}$ as 5. When you can do that, you should: 10 is the simplified form of $\sqrt{100}$, and 5 is the simplified form of $\sqrt{25}$. Make sure that when a question asks you to simplify your answer, you don't leave your answer in a form like $\sqrt{25}$. Even though the value is exactly the same as 5, you might be marked wrong or lose some points, because you haven't followed the directions! In math, a problem is generally not considered complete until the answer is written in the most simplified form, unless the directions explicitly tell you not to simplify.

Don't leave any perfect squares as factors under the radical! If there's a perfect square under there that you know the root of, you should bring it outside of the radical sign by taking the square root of it.

Now you know how to simplify when there's a perfect square under the square root symbol. And you know that some radicals cannot be simplified: for example, $\sqrt{2}$ cannot be simplified further, so it is fine to leave $\sqrt{2}$ in your answer. But what if you're asked to simplify something like $\sqrt{50}$? You don't know the square root of 50 off the top of your head—it's not a perfect square. You do know, by estimating, that the answer will be a little bigger than 7 by estimating, but you're not quite sure if you can further simplify. It turns out that you can! Just follow all the rules you know about roots and multiplication and see how else you can write $\sqrt{50}$.

For starters, you can write the number 50 in whatever form you want. So, if you wanted, you could rewrite $\sqrt{50}$ as $\sqrt{(25\times2)}$. All you've done is replace the number 50 with another way of saying the number 50; you haven't changed the value of the expression at all. And now, because you have multiplication under the square root sign, you know that it's okay if you break the expression up and change $\sqrt{(25\times2)}$ to $\sqrt{25}\times\sqrt{2}$. You can do that because you just proved that when you multiply two square roots, it's the same as just multiplying the values under the square roots and then taking the square root of their product. But why would you want to go through all that? Well, because *now* you have something you know how to simplify!

You know that $\sqrt{25}=5$, so you can rewrite $\sqrt{25}\times\sqrt{2}$ as $5\times\sqrt{2}$. Now you're finished. There's nothing more here that can be simplified, because $\sqrt{2}$ is as simple as it can get; there are no more perfect squares that you can factor out. Conventionally, you could rewrite $5\times\sqrt{2}$ as $5\sqrt{2}$. Those two expressions mean exactly the same thing and have the exact same value, but the second format is usually preferred because it's easier to read. In the end, if you were asked to simplify $\sqrt{50}$, the correct answer would be $5\sqrt{2}$.

What's the process? Take a look at each step. For an example, let's say that you're asked to simplify $\sqrt{48}$.

1. Look at the number underneath the radical and try to think of a perfect square that is a factor of that number. (Ignore the perfect square 1, because factoring out a 1 doesn't do anything.) If you can't find a perfect square by looking, use a factor tree to help you. In this case, you can tell by looking that 48 has 4 as a factor.
2. Rewrite the number under the radical so that the perfect square you found is factored out. When you're finished, there should be a multiplication

expression underneath the radical. In this example, you want to factor out a 4, so you're going to rewrite the expression as $\sqrt{(4\times12)}$.

3. Split up the radical so that you have two radicals being multiplied together. In this example, you would rewrite the problem as $\sqrt{4}\times\sqrt{12}$. (By the way, another way to write $\sqrt{4}\times\sqrt{12}$ is $\sqrt{4}\sqrt{12}$. If the multiplication sign helps you for now, keep it until the very end.)
4. Simplify the radical expression that includes the perfect square. Because $\sqrt{4}=2$, you can rewrite $\sqrt{4}\sqrt{12}$ as $2\sqrt{12}$.
5. Repeat Steps 1–3 with the remaining radical. If any perfect square factors still remain, they must be factored out. In this example, take a look at $2\sqrt{12}$ and check the number underneath the radical. You'll see that it does indeed have a perfect square as a factor (there's another 4 in there!). So you have to repeat the process. The 2 out in front of the radical is fine; you can leave it alone. But you're going to rewrite the whole expression as $2\sqrt{(4\times3)}$. Because the numbers under the radical are being multiplied, you can take the root of each of them separately, rewriting the expression as $2\sqrt{4}\sqrt{3}$. At this point, you can simplify the radical because it has a perfect square under it, leaving you with $2\times2\times\sqrt{3}$.
6. Multiply any integers together and remove the multiplication sign from your answer. In this example, you will change $2\times2\times\sqrt{3}$ to $4\times\sqrt{3}$, and then drop the multiplication sign to get a final answer of $4\sqrt{3}$.

You can take a nice shortcut here if you understand where it's coming from. Any time the number under the radical has a perfect square as a factor, you can factor out that perfect square, take the square root of it, and write that square root outside of the radical.

Notice that you could have saved yourself the repeated steps in the middle if you'd realized that 48 could have been rewritten as 16×3. Sixteen is the biggest perfect square that's a factor of 48, which is why your final answer has the square root of 16 factored out of it. But you didn't spot the 16 right away, you just had to go through an extra step to get there. It's totally fine to do your factoring one perfect square at a time when simplifying radicals.

Raising Exponents to Exponents

Exponents are a whole new code to learn, so they can be sort of intimidating at first. The most intimidating-looking exponent problems have exponents raised to other exponents! For example, you might be asked to simplify something like $\left(8^2\right)^3$. Yikes! But don't worry. If you take a second to think about what this code actually *means*, it should be pretty clear what you're supposed to do.

Since $\left(8^2\right)$ is in parentheses, your order of operations tells you that you're supposed to perform the operation first. You know that $\left(8^2\right)=64$, but that's not going to help you figure out a shortcut, so for now you're just going to rewrite it as (8×8). When you plug that into the original expression, you now have $(8\times8)^3$. There are two ways you can think about this. One way is to say that anything raised to the third power must be multiplied as a factor three times, so $(8\times8)^3=(8\times8)(8\times8)(8\times8)$. Now you can see that this expression is six 8s multiplied together, which you can rewrite as 8^6. Or, if you prefer, you could go back to the $(8\times8)^3$ step and, because you know how multiplying with exponents works, rewrite this as $(8^3)(8^3)$. Since you know the rules of multiplying with exponents, you can simplify that expression to $8^{(3+3)}$, or 8^6. Either way, you're doing the same thing and getting the same result.

What's the Shortcut?

Well, when you simplify $\left(8^2\right)^3$, you end up with 8^6. See what happens? The exponents just get multiplied together. That's because however many times you multiply the base together in the bottom exponent, you repeat that process however many times the top exponent tells you to. Knowing this rule makes questions that have an exponent raised to an exponent much easier! For example, without writing out any work, you can now simplify $(5^2)^7$ to 5^{14}, or you can simplify $(5^2 3^2)^2$ to $5^4 3^4$. When in doubt, write it out! $(5^2 3^2)^2$ is equal to $(5\times5\times3\times3)^2$ which is equal to $(5\times5\times3\times3)(5\times5\times3\times3)$. That's four fives being multiplied by four 3s, so the final answer has to be $5^4 3^4$.

You can use this rule to help you make the bases of exponential equations the same. You're probably beginning to understand why getting comfortable with multiplication tables and factoring is such an important skill to have for pre-algebra and beyond. It really helps if you are able to look at numbers and notice the factors they have in common. For example, let's say that you're asked to simplify the expression $4^6 2^{15}$. Well, if you had all the time in the world, you could do this multiplication by hand. But, not only would it take you a long time and give you an unmanageable big number; it also wouldn't be what your teacher was looking for as the answer. Instead, you need to find a simpler way to write the expression using exponents.

If the bases of these numbers were the same, you could multiply them by adding the exponents. If the bases aren't the same, there's really not much you can do. Oh! But if you could *make* the bases the same, then you can do it! When you look back at the expression $4^6 2^{15}$, you'll see that you could change 4 so that it has 2 as a base. In fact, you know that $4 = 2^2$, so you can replace 4 with 2^2 and the value shouldn't change at all. Once you do that, you have $(2^2)^6 2^{15}$. Now, this you can do! You know that when you raise an exponential equation to another exponent, the exponents get multiplied. So you can rewrite the expression as $2^{12} 2^{15}$. And now you know that when you multiply two exponential equations with the same base, you add the exponents. Therefore, you can simplify $2^{12} 2^{15}$ and rewrite it as 2^{27}. Done!

The trick here is going one step at a time, and using the rules of exponents that you've learned so far. If you get stuck or confused, you can either write out the expression without exponents or just think of it without exponents. For example, if you think about 4^6, you might realize that you can visualize it as six 4s being multiplied together, like this: $4 \times 4 \times 4 \times 4 \times 4 \times 4$. If you were to replace each of those fours with 2×2, you'd have a string of twelve 2s being multiplied together. That's the reason that $4^6 = 2^{12}$. Or, maybe you don't know how to rewrite 4^6 right away, but you know you want the base to be 2, so you change it to $(2 \times 2)^6$. Now you can follow the rules of exponents and change $(2 \times 2)^6$ into $2^6 2^6$, which you know is equal to 2^{12} because when you multiply exponential expressions with the same base, you add the exponents. No matter how you do it, as long as you follow the rules of math, you end up with the same answer. So if you can't see the path right away, just take it one step at a time!

One Last Example

Take a look at one last example of the raising-exponents-to-exponents rule. What if you're asked to simplify $15^4 30^3$? Okay, admit it: this expression looks pretty hard. But you just have to look for one thing you know how to do at a time. You're going to start by rewriting that 30 as 15×2, because you can see that doing so will at least let you put some of the bases together. This way, $15^4 30^3$ becomes $15^4(15 \times 2)^3$. Now use your multiplication rules to distribute that exponent to both the terms in the parentheses, which will let you rewrite the expression as $15^4 15^3 2^3$. At this point, you have two exponential expressions with the same base being multiplied together, so you can add their exponents to get $15^7 2^3$. That looks pretty good! You can't factor any perfect squares out of the 2 or the 15, and they don't have anything in common, so you're finished.

You might remember from a few chapters ago something called prime notation. Prime notation is a way to show a number where the bases are the prime factors of that number, and the exponents count how many times those primes show up as factors. Let's say you're asked to write $15^4 30^3$ in prime notation. Now you have to take a look at your result, $15^7 2^3$, and see if there are any bases in the problem that aren't broken down into primes. 2 is okay, because it is prime, but 15 is not, so you have to rewrite it as the product of its prime factors. You're going to change the expression $15^7 2^3$ to read $(5 \times 3)^7 2^3$. Now all your bases are prime, and you just have to get an exponent next to each one. Therefore, you're going to distribute that 7 based on the rules of multiplication with exponents, and now the expression will read $5^7 \times 3^7 \times 2^5$. For your final step, you take out the multiplication signs and write your prime notation as $2^5 3^7 5^7$. All the bases are different primes, and they each have their own exponent, so now you're *really* finished!

Another Path

Try going through this example using another path. If the directions for a problem ask you to write $15^4 30^3$ in prime notation, what other path could you follow? Well, you could start by factoring the bases and writing them as primes. If you did that, you would rewrite the expression as $(3 \times 5)^4(2 \times 3 \times 5)^3$. Now, following the exponent rules of multiplication, you can distribute those exponents to each number being multiplied in the parentheses and rewrite $(3 \times 5)^4(2 \times 3 \times 5)^3$ as $3^4 5^4 2^3 3^3 5^3$. Looks good so far! Start by putting the numbers with the same bases next to one another and rearranging the expression to look like $2^3 3^4 3^3 5^4 5^3$. Now, for the final step, you've got to simplify by combining the bases by adding their exponents, which leaves you with $2^5 3^7 5^7$. That's the same thing you got before!

The road was slightly more direct this time since you knew you wanted prime notation from the beginning, but you ended up in the same place.

Adding and Subtracting with Exponents and Roots

You already know how to multiply, divide, factor, and simplify exponents and roots, raise them to other exponents, and put them in prime notation—whew!—but what about adding and subtracting exponents and roots?

You can't really add or subtract exponents and roots the same way you do with other types of numbers. There's no nice shortcut like there is for multiplication and division and raising powers to other powers. Why? Because at their core, exponents and roots are multiplication problems. They tell you what happens when you take a number and multiply it by itself, so they combine really well with other rules related to multiplication, such as multiplication itself, division, and other exponents. But they don't combine very well with addition and subtraction.

Take a Look

What if you're asked to simplify 3^2+3^3? It certainly looks like there would be some shortcut, but there isn't. When it comes to addition and subtraction with exponents, your options are limited. In this case, the numbers are small enough that you can figure them out and rewrite this problem as $9+27$, which is 36.

If the numbers are too big to write out, then you really can't simplify at all. For example, let's say you're asked to simplify $2^{10}+3^{16}$. You can't. It's already as simple as it's going to get. The only way you could possibly change it would be to calculate each of the values and add them together, and you're most likely not going to be asked to do that with such big numbers.

There is one thing you can do with exponents and roots when you're adding them together or subtracting them from one another. If you're adding or subtracting the exact same term, you can treat those two identical terms as units. Say, for example, that you're asked to simplify 3^5+3^5. Well, without writing these numbers out, you really don't have a shortcut for adding them. But wait a second: the two terms are exactly the same! Even if you don't know what 3^5 equals, you know that if you add another 3^5 to it, you'll have two of them. So just as "an orange plus an orange equals two oranges," it's also true that "a 3^5 plus a 3^5 equals two 3^5s." In math terms, $3^5 + 3^5 = 2(3^5)$. Any time you add something to itself, you end up with two of that thing. It helps to think of "a 3^5" as an item, like "an apple" or "an orange." You don't have a mathematical value for an orange, but you can still tell when you have two oranges. So when you add two of any value together, you'll get two times that value.

The same addition and subtraction rules hold true for square roots. If you're asked to simplify $\sqrt{2}+\sqrt{3}$, you just can't do it. There's no way you can combine that information any more than it's already combined. But, if you're asked to simplify $\sqrt{2}+\sqrt{2}+\sqrt{2}$, there's one thing you can do. First, you can see that the terms are the same, and you're adding three of them together, so your final result should be three times the value of that term. So just like an orange plus an orange plus an orange would give you three oranges, a $\sqrt{2}$ plus a $\sqrt{2}$ plus a $\sqrt{2}$ will give you three $\sqrt{2}$ s, or $3\sqrt{2}$. And the same rule holds true for subtraction. For example, $\sqrt{10}-\sqrt{10}=0$. You don't know what $\sqrt{10}$ is, but you know that if you take it away from itself, you'll have nothing left.

A Few More Examples

Take a look at some more examples. What if you're asked to simplify $3^2+3^2+3^2+3^2$? Well, it's addition with exponents, so at first you should be very cautious. Notice, though, that all the terms are the same. The term is 3^2, and you're adding up four of them, so when you're finished you'll have four of those 3^2 terms. In other words, you'll have $4(3^2)$. If you wanted, you could solve this problem now: $3^2=9$, so $4(3^2)=36$. (If you want to double-check this idea, you can write out $3^2+3^2+3^2+3^2$, and you'll get the same answer.) If you wanted or were asked to, you could also write $4(3^2)$ in prime notation. The 3 is already prime, but you'd have to break down the 4 and rewrite $4(3^2)$ as $2^2 3^2$.

Now, what if you were asked to simplify $3^2+3^2+3^2$? You have three of the same term, so when you add them, the result should end up as three times that term, or $3^2+3^2+3^2=3(3^2)$. But this time, something cool has happened. You can actually further simplify your answer, because you're multiplying two numbers that have the same base. Both 3 and 3^2 have a base of 3, so you can multiply them by adding their exponents. But wait! Three doesn't have an exponent. What do you do? Well, every number that doesn't have an exponent really has an exponent of 1, which is the default, so you can rewrite $3(3^2)$ as $3^1 3^2$. Now you're multiplying two numbers with the same base, so you add the exponents and simplify this expression to 3^3. You might not have guessed at first glance that $3^2 + 3^2 + 3^2 = 3^3$, but now you know that it does. And you can prove that you were right! You know all these numbers, so you can rewrite this equation as $9 + 9 + 9 = 27$, which you know is true.

Chapter 9 Exercises

Simplify the following expressions. Write your answer as an exponential expression.

1. $3^3 \times 3^3$
2. $2^5 2^6$
3. $5^9 \times 5$
4. $7^2 \times 7^0$
5. $11^4 11^7$
6. $3^3 \times 3^2$

Simplify the following expressions. Write your answers as an exponential expression.

1. $3^3 \div 3^3$
2. $2^6 \div 2^5$
3. $\frac{5^9}{5}$
4. $\frac{7^2}{7^0}$
5. $11^7 \div 11^4$
6. $3^3 \div 3^0$

Simplify the following expressions. Write your answer as an exponential expression.

1. $\sqrt{4}\sqrt{4}$
2. $\sqrt{10} \times \sqrt{10}$
3. $\sqrt{10} \times \sqrt{10} \times \sqrt{10}$
4. $\sqrt{8} \times \sqrt{2}$
5. $\sqrt{8} \div \sqrt{2}$
6. $\sqrt{7} \div \sqrt{7}$
7. $\dfrac{\sqrt{27}}{\sqrt{3}}$
8. $\sqrt{5}\sqrt{20}$

Simplify the following radicals as much as possible.

1. $\sqrt{100}$
2. $\sqrt{20}$
3. $\sqrt{45}$
4. $\sqrt{2}$
5. $\sqrt{1}$
6. $\sqrt{80}$
7. $\sqrt{128}$
8. $\sqrt{75}$

base of 3, so you can multiply them by adding their exponents. But wait! Three doesn't have an exponent. What do you do? Well, every number that doesn't have an exponent really has an exponent of 1, which is the default, so you can rewrite $3(3^2)$ as $3^1 3^2$. Now you're multiplying two numbers with the same base, so you add the exponents and simplify this expression to 3^3. You might not have guessed at first glance that $3^2+3^2+3^2=3^3$, but now you know that it does. And you can prove that you were right! You know all these numbers, so you can rewrite this equation as $9+9+9=27$, which you know is true.

Chapter 9 Exercises

Simplify the following expressions. Write your answer as an exponential expression.

1. $3^3 \times 3^3$
2. $2^5 2^6$
3. $5^9 \times 5$
4. $7^2 \times 7^0$
5. $11^4 11^7$
6. $3^3 \times 3^2$

Simplify the following expressions. Write your answers as an exponential expression.

1. $3^3 \div 3^3$
2. $2^6 \div 2^5$
3. $\frac{5^9}{5}$
4. $\frac{7^2}{7^0}$
5. $11^7 \div 11^4$
6. $3^3 \div 3^0$

Simplify the following expressions. Write your answer as an exponential expression.

1. $\sqrt{4}\sqrt{4}$
2. $\sqrt{10}\times\sqrt{10}$
3. $\sqrt{10}\times\sqrt{10}\times\sqrt{10}$
4. $\sqrt{8}\times\sqrt{2}$
5. $\sqrt{8}\div\sqrt{2}$
6. $\sqrt{7}\div\sqrt{7}$
7. $\dfrac{\sqrt{27}}{\sqrt{3}}$
8. $\sqrt{5}\sqrt{20}$

Simplify the following radicals as much as possible.

1. $\sqrt{100}$
2. $\sqrt{20}$
3. $\sqrt{45}$
4. $\sqrt{2}$
5. $\sqrt{1}$
6. $\sqrt{80}$
7. $\sqrt{128}$
8. $\sqrt{75}$

Rewrite the following expressions in prime notation.

1. 4^2
2. $(2^3)^4$
3. $(25^3)^3$
4. 12^5
5. 100^3
6. $(36^2)^3$

CHAPTER 10

Base-Ten System

When you talk about the base-ten system, you have to include fractions, exponents, addition and subtraction, multiplication and division, and pretty much everything else you've read about so far. This chapter is a chance for you to review the base-ten system: the system upon which number writing is based. The base-ten system is what gives meaning to the place values and decimals of numbers, and it's something you had a lot of experience with back when you were learning to add or subtract two-digit numbers and three-digit numbers.

What Is the Base-Ten System?

There are a lot of different systems of numbers. For example, the system of numbers used in computer programming is called binary, and the whole system is in base two. When counting time, you use a system based on the number 60—every time you get to 60 seconds, you call it a minute; every time you get to 60 minutes, you call it an hour. The number system you use for pretty much everything else, and certainly everything you'll do in algebra, is in base ten. This means that all math students have a way of writing numbers that is based on groups of ten.

The units digit, sometimes called the ones digit, starts off by telling you how many of something you have. Once you get to ten things, you group them together into a group of ten and show that with your tens digit. The number 13, for example, has a 1 and a 3 in it, but those numbers mean different things. The 3 tells you how many groups of one you have, and the 1 tells you how many groups of ten you have. With one group of 10 and three groups of 1, you have a total of 13 items.

The good news is you've been using the base-ten system your whole life, so it already makes sense to you. It seems obvious and is almost harder to explain than it is to use. You can glance at a number like 165 and recognize instantly what it means, because you're so used to the code.

The *units digit* and the *ones digit* are just two different terms for the same thing. The *units digit* is the term by which you understand everything else—move to the left, and you get groups of ten units. Move to the right, and you get groups of tenths of a unit.

Place Value

Place value is the term for the different digits of a number. You have already learned about the **units digit**, or ones digit. For the most part, you're going to call it the units digit because that's the more conventional name, but if you or your teacher or textbook call it the ones digit, that's fine. The two terms mean the exact same thing.

The units digit tells you how many "ones," or single counting units, are included in the number. In any number, there is space for only a single digit in each place value. Because of this, once you get to ten units, you group the single units together and call them one unit of ten. The value of that group of ten is shown in the **tens digit** of a number.

As an example, let's say you're adding together 8 and 7. When you add these groups of eight units and seven units, you can make one full group of ten, with five units left over. Thus you have one group of ten and one group of five. This means your tens digit, which tells you how many tens you have, is 1. It also means that your units digit, which tells you how many single units you have that cannot be grouped into bigger groups, is 5. That's why when you add two numbers together, you have to carry over to the bigger digits. Any time you add together two big numbers on paper, you do this by "carrying over" amounts to the next column of numbers as you add.

It's important for you to know the name of each place value. Use the number 1,234,567.890 as an example. The digit directly to the left of the decimal point is the units digit. The units digit here is 7. Moving toward the left, each number counts values that are ten times bigger than the place value next to it. Therefore, 6 is the tens digit, telling you how many tens there are in this number. 5 is the hundreds digit; 4 is the thousands digit; 3 is the ten-thousands digit; 2 is the hundred-thousands digit; and 1 is the millions digit. You could go on and on forever, with the names of the place values getting ten times bigger every time you move one place to the left.

Then, to the right of the decimal point, the values get ten times smaller each time you move to the right. So the 8 in this example is the tenths digit; the 9 is the hundredths digit; and the 0 is the thousandths digit. Notice that the names of the place values to the right of the units digit have a "th" sound in them, and the names to the left of the units digit don't.

Read carefully when you're asked to round to a certain digit! The words *hundreds* and *hundredths* look very similar, but they're not the same at all! In the number 234.567, for example, the *hundreds* digit is 2, but the *hundredths* digit is 6.

Rounding

You probably already learned about rounding numbers before you began your study of pre-algebra. As you may have noticed, pre-algebra is kind of a chance to review all the rules you've learned about math so far, learn a little more about them, and see how they all fit together before you start adding variables into the mix.

When you're asked to "round" a number in math, you're asked to write down an estimate of its value. Almost always, the directions in a problem will tell you what place value you should round to. For example, if the directions tell you to "round to the nearest whole number," "round to the nearest integer," "round to the nearest units digit," or "round to the nearest ones digit" (can you believe you have that many ways to say the same thing?), they're telling you to round whatever is after the decimal point so that you can express your answer without the decimal.

So let's say you're asked to round 4.5980 to the nearest whole number. This number is somewhere between 4 and 5, so one of those will be the nearest whole number. Which one is actually nearer? 4.5980 is closer to 5 than 4 because it's equal to or bigger than 4.5, so you're going to round up to 5.

In short, when you're asked to round, here are the steps you should follow, using "4.5980 rounded to the nearest integer" as an example.

1. Find the place value you are expected to round your answer to. This problem is asking you to round to the nearest integer, which means it must end before the decimal point, so you're being asked to round to the nearest units digit.
2. Move one place value to the right of that to find the value you will use to determine the direction of your rounding. In this case, 5 is the value in the place value directly to the right of the place value you want to end your number with.
3. If the value from Step 2 is 5, 6, 7, 8, or 9, round the value from Step 1 up one digit. If the value from Step 2 is 0, 1, 2, 3, or 4, leave the value from Step 1 just as it is.
4. Replace all values to the right of the value in Step 1 with the number zero. If the zeroes are unnecessary, ignore this step.

Take a look at two more examples to make sure you know what to do. What if you're asked to round 4,562 to the nearest hundred? Well, the question is about hundreds, so you locate the hundreds digit, which is 5. Then, you look one place to the right at the tens digit, which is 6. Because that digit is 6 (greater than or equal to 5), you have to round your hundreds digit up one value, so it goes from 5 to 6. Now you replace all the remaining digits with zeroes, and you know that 4,562 rounded to the nearest hundred is 4,600. This should make sense. Think about 4,562—it's closer to 4,600 than 4,500, so 4,600 is the closest multiple of 100 to 4,562. What if you're asked to round 3.4562 to the nearest hundredth? The hundredths digit is 5. When you look directly to the right at the thousandths digit, you'll see the value is 6, which means you need to round up. Thus, 3.4562 rounded to the nearest hundredth is 3.46.

Expanded Form

Expanded form is a way of writing a number as a sum of its units, tens, hundreds, and other groups represented by digits in the base-ten system. In other words, it's a way of writing a number as an expression that tells you how many ones, tens, hundreds, thousands, or any other place value are in that number.

For example, if you wanted to write the number 482 in expanded form, you would take a look at the different digits and get your information from there. The number 482 has a 2 in the units place, so it has two individual units. It has an 8 in the tens place, so it must have eight groups of ten, which is a total of eighty units. It has a 4 as the hundreds digit, which means it has four groups of 100, which is a total of 400 units. So the number 482 is actually $400 + 80 + 2$.

That's all expanded form is. It's mostly used as a way of helping you understand what the place values mean. You might be asked to prove you can write numbers in expanded form on a quiz or test, but you probably won't use it very much apart from that.

Multiplying and Dividing by Powers of Ten

You already know how to multiply and divide by ten, and thanks to your practice with exponent rules, you also know how to multiply or divide by ten when it's raised to any exponent. But there's a nice shortcut that will make your life easier when it comes to multiplying or dividing by ten or by ten raised to any power. And it's always good to have shortcuts!

You've probably noticed that commas are often used in numbers that are more than three digits long in order to make them easier to read. The commas don't have any mathematical meaning; 1234 has exactly the same mathematical meaning as 1,234.

This shortcut works because you are dealing with a system based on powers of ten. Each place value represents the number ten raised to a different exponent. Take a look at the number 1,234,567,897. In this example, 7 is the units digit, so it does not represent any groups of ten. Because of that, it could be expressed as "seven groups of no tens," which you would write mathematically as 7×10^0 (because, remember, anything raised to the zero power is equal to 1). $7 \times 10^0 = 7$. The tens digit, which in this number is 9, tells you how many groups of ten you have. It tells you that you have nine groups of ten, which you can show mathematically as 9×10^1. Notice that this is exactly what a 9 as the tens digit tells you. The number 9 in the tens place contributes a value of 90 to the number, as you would see if you wrote the number out in extended form.

Why can you move the decimal point as a way of multiplying or dividing by tens?
Because numbers are in a base-ten system, every place value is ten times bigger than the one to the right of it. When you move the decimal point one place to the right, you're making the number ten times bigger. When you move it one place to the left, you're making the number one tenth of the size.

Do you see what's happening? You can represent the value of each place value as ten raised to a power. The units digit tells you how many 10^0 there are. The tens digit tells you how many 10^1 you have. The hundreds digit tells you how many 10^2s you have, which all should make sense because 10^0 is equal to 1, 10^1 is equal to 10, 10^2 is equal to 100, and so on and so on.

What's the point of all this? The point is that you can trade place values for powers of ten. When you multiply a number by ten, making it ten times bigger, it's like adding a decimal place to the number. You are already used to doing this with integers. You know that 43×10, for example, is 430. And you know that $13 \times 100 = 1,300$. When you multiply 43×10 to get 430, you're really just moving the decimal point one place to the right by adding a place value. When you multiply $13 \times 100 = 1,300$, you're just adding two decimal places onto the end of the number.

Now you can take that same rationale and use it when you're multiplying or dividing with decimals. For every power of ten you multiply by, making the number ten times bigger, you can move the decimal point one place value to the right. For every power of ten you divide by, making the number ten times smaller, you can move the decimal point one place value to the left.

More Examples

Take a look at a couple examples to see how this might help you. Want to figure out 4.89×10^3? Well, $10^3 = 1,000$, because every power of ten just adds another zero to the end of the number. So when you want to simplify 4.89×10^3, you're multiplying by 10 three times, which moves the decimal point three places to the left. Doing so leaves you with 4,890.

What if you're asked to find 235.12×10^4? You're multiplying by a power of ten, so you just need to move the decimal point four places to the right. That means $235.12 \times 10^4 = 2,351,200$. The same thing works in reverse for division. If you're asked to find $4,318 \div 10^3$, no problem. When you divide by a power of ten, you can just move the decimal point to the left however many times the exponent tells you to. So $4,318 \div 10^3 = 4.318$. If you wanted to divide $71 \div 10^4$, what would happen? For starters, when you try to move the decimal point to the left, you'll realize you don't have four places to move it. When this happens, you can add zeroes as placeholders for the empty decimal places. If it helps, you can think of the fact that, although you'd never write this, 71 has the same value as 0071. Because of that, you can add all the placeholder zeroes you need.

Density Property of Numbers

One property you should know that is related to the idea of fractions and decimals is the density property of numbers. The **density property of numbers** tells you that between any two numbers on the number line, there is always another number. For example, between 0 and 1, there's ½. Between 0 and ½, there's ¼. Between 0 and ¼, there's ⅛. And you can keep going and going and going.

This might be easiest to see with decimals. Even when you take two numbers that seem very, very close together, there's always another number between them. For example, think about .00000001 and .00000002. These are certainly very close. But between them you have an infinite list of numbers. For example, .000000015 is between these two numbers.

Chapter 10 Exercises

1. Round 34.56 to the nearest integer.
2. Round 111.111 to the nearest tenth.
3. Round 99.01 to the nearest hundred.
4. Round 123.123123 to the nearest ten thousandth.
5. Round 54.1 to the nearest ten.
6. Round 4,329.29 to the nearest tenth.
7. Round 94.4 to the nearest whole number.
8. Round 4,499 to the nearest thousand.
9. Round 35.51 to the nearest units digit.
10. Round 41.190 to the nearest tenth.
11. Round 41.190 to the nearest hundredth.

Simplify the following and express as one term in decimal form.

1. 4.56×10^3
2. $1,200\div10^2$
3. $\frac{4,000}{10^3}$
4. $\frac{250}{10^5}$
5. $\frac{4,000}{40}$
6. $\frac{160,000}{400}$
7. $.000134\times10^7$
8. $1.2\times1,000$
9. $\frac{34.07}{1000}$
10. $.0023\times100$

CHAPTER 11

Simplifying Expressions and Equations (Without Variables)

In this chapter, you'll start to get a feel for what an expression is and what you're allowed to do to it. You're going to work with just numbers for now, because you've been studying numbers for years and you understand a lot about them. The idea is to use numbers to demonstrate the rules, so that you can apply the same rules when you start adding variables.

What's an Expression?

Most of the problems you've seen in math so far are just expressions. For example, when a problem asks you to find 4×2, it is asking you to simplify the expression. An expression is just some math grouped together, and a problem is usually asking you to combine it or write it in a simpler way. This is called **simplifying** an expression—putting it into the simplest format. Of course, 8 means the same thing as 4×2, but as one number, it's easier to read. It has had all the operations removed and has been simplified so that it can be expressed as one number.

When you find 4×2, you aren't actually solving for anything. There's nothing to solve for. There's no mystery. You're just simplifying, writing the numbers in a new form. Expressions don't have an equal sign. They are just different ways of expressing a number, and you can change the way a number is expressed as long as you don't change the value of the number.

There is a lot of new terminology in pre-algebra. If you ever get confused about any terminology or aren't sure what's being asked, ask for help! You can't get the right answer if you don't understand what the question is.

So for example, if you want, you could change 4×2 into 8. Or you could change it into $6+2$. Or you could change it into $\frac{16}{2}$, or $10-2$, or $\frac{800}{100}$, or 1×8. You know these expressions are equal because, after you perform the mathematical operations, they simplify to the same number, which is 8 in this case.

Order of Operations

Think about all the ways that you could express the number 8. They are all equal to 8, but they look pretty different. Because of this, you have to have a set of rules that tells you how to perform mathematical operations on all the different ways that numbers can be written.

Math exists in the real world: you can count things; you can add things; you can multiply and divide things, and so on. But written math is something that someone invented. It's just a system for writing down true information so that it stays true when you manipulate it.

One of the most important rules to master is the **order of operations**. Operations are actions like adding, subtracting, multiplying, and dividing. Operations have to be done in a certain order to make the math come out correctly. Math is basically a code for representing information, so you have to follow the code, or you'll end up at the wrong answer.

Examples of Order of Operations

Many people remember the order of operations by using the mnemonic **PEMDAS**. (A mnemonic, pronounced "nee-MON-ick," is a trick that helps you remember something.) The letters in PEMDAS stand for:

Parentheses
Exponents and roots
Multiplication and **D**ivision
Addition and **S**ubtraction

The order of operations is imposed to ensure that everyone gets the same result when working with an expression that has more than one step. For example, take a look at the expression $5+3\times4$. Without a clear set of rules, some people might start by adding 5 and 3, which would result in their interpreting the expression as having a value of 32. Others might start by multiplying 3×4 and determine that the value of the expression is 17. But only one of those can be right! So you need a set of rules that tells you the right place to start. In fact, even though the addition comes first in the expression, PEMDAS says you have to do multiplication before addition, so this expression does have a value of 17.

Let's look at a more complex example:

$$2(4+2)^2-3(9)+\frac{10}{2}$$

First, look for any **parentheses** and handle what happens in those. Now you can rewrite the expression as:

$$2(6)^2-3(9)+\frac{10}{2}$$

The next step is to deal with any **exponents or roots**. If you didn't have any, you could just skip this step. But you do! So you have to apply the exponent next and then see where you are.

$$2(36)-3(9)+\frac{10}{2}$$

Now look for any **multiplication and division** and do them in order from left to right. Once you've done that, the expression looks like $72-27+5$. And now you have to do the **addition and subtraction** from left to right as well. So first you'll subtract the 27, and then you'll add the 5. Your final step of simplification leaves you with 50.

Note that you haven't actually solved anything. There wasn't ever any question to answer! You were given one way of writing a number, and you were just supposed to write it in a simpler and better way. Now you're finished.

Be *very careful* when it comes to the order of operations. When division and subtraction are involved, you have to make sure you follow the rules exactly, doing all multiplication and division from left to right and then doing all addition and subtraction from left to right.

It is *very important* to follow this order. Say you didn't know the order of operations. When you looked at this expression, you might have wanted to start by multiplying 2 by 4, since they are next to each other and one is in parentheses. But if you just went left to right doing one operation at a time, you'd end up in the wrong place. Check it out: if you *mistakenly* look at the expression $2(4+2)^2-3(9)+\frac{10}{2}$ and think that you can just do the operations as they come along from left to right, you're going to get the wrong answer, because you're reading the code wrong. If you were to ignore PEMDAS and just do all the operations from left to right, you would get 441.5 as your answer—and that's *really* far off from where you want to be. All because you misread the code and did some illegal math. For math to be a good code, it has to spit out the same simplification or solution every time. PEMDAS is a convention that must be honored because it makes sure you are reading math expressions correctly *and* are writing them in such a way that others will read them as you intended.

Commutative Property

Up until now, you haven't had to do much in math other than simplify expressions. And there's probably been very little reason to move the numbers around or write them in new ways. But now that you're getting ready for algebra, it's important to make sure you understand a few properties that numbers—and operations on numbers—have. The good news is, you already know all these properties and know that they're always true. You just might not know what they're called.

The commutative property has to do with the order of numbers in relation to their operation. To tell whether an operation is commutative, you just have to figure out if changing the order of the operation changes the result of the operation.

For example, check out the following equations:

$$2+3=3+2$$
$$1+9=9+1$$
$$-3+3=3+-3$$

These are all equations, because they have an equal sign in them. You know that all of these are true statements. Once you simplify them, you'll realize that these equations basically say $5=5$, $10=10$, and $0=0$. That's your proof that addition is commutative. The order of the numbers doesn't matter if all you're doing is adding them.

Because of the commutative property of addition, you know that, for example:

$$1+2+3+4+5+6=1+6+2+5+3+4$$

You can put the pieces you are adding up in whatever order you want, and the final value of the sum of the terms won't change. Therefore, addition is commutative.

And so is multiplication! For all the same reasons. Here are some examples:

$$2\times3=3\times2$$
$$1\cdot9=9\cdot1$$
$$(-3)(3)=(3)(-3)$$

These equations are true because $6=6$, $9=9$, and $-9=-9$. Therefore, multiplication is commutative. It doesn't matter what order you multiply the numbers in; the final simplification will come out to be the same.

You can remember that this property is called the *commutative* property because it has to do with moving the numbers around. Like when you *commute* to school, you are riding there and back, when you *commute* the numbers, you move them around to a new place.

Subtraction and division are *not* commutative. In these operations, the order does matter! For example:

$$3-7 \neq 7-3 \text{ (because } -4 \neq 4)$$
$$\frac{8}{2} \neq \frac{2}{8} \text{ (because } 4 \neq \frac{1}{4})$$

So you have a **commutative property of addition** and a **commutative property of multiplication**, but you don't have a commutative property of subtraction or division. It seems silly to learn this property with numbers, because you'd never take the time to write 1×9 as 9×1. But once you get into the *x*s and *y*s of algebra, you might have a reason to write things differently.

Associative Property

The associative property applies when something can be simplified with the same result no matter how you group the numbers together. In other words, you can add parentheses anywhere you want, and it won't make a difference.

Here's how it works. If you simplify $2+(3+5)$, does it come out to be the same as $(2+3)+5$? Go ahead and check. Is $2+8$ equal to $5+5$? Yes, they're both equal to 10. So it must not matter which way you group things when you're adding them together. Here are some more examples:

$$(9+0+3)+1=9+0+(3+1)$$
$$-2+(9+7)=(-2+9)+7$$

That's why you have an **associative property of addition**. If you're adding up a bunch of numbers, you can add them in whatever groups you want, and the net effect will still be the same. Think about taking 100 pieces of candy from a box and putting them in a pile. You can put them in one at a time, or two at a time, or five at a time, or fifty at a time, and it won't matter. Once you add all 100 of those pieces of candy to the pile, you'll have 100 pieces of candy in the pile.

You can remember that this is called the *associative* property because it has to do with how the numbers *associate.* When you associate with your friends, you are grouping together with them. Similarly, the associative property has to do with how the numbers are grouped together.

There is also an **associative property of multiplication**. You can multiply numbers in whatever groups you want, and the simplification will be the same. Check out the math below so you can confirm that it works:

$$1 \times 4 \times (2 \times 2) = (1 \times 4 \times 2) \times 2$$
$$(2 \times 3) \times 5 = 2 \times (3 \times 5)$$

Multiplication comes out to be the same no matter where you put the parentheses. If you start with $10 and then double your money (or multiply it by 2), and then triple your money (or multiply it by 3), you go from $10 to $20 to $60. If you start with $10 and then make it six times as big (multiply it by 6), you still end up with $60. It doesn't matter whether you do all the multiplication at once or one at a time—you still get all the benefits by the time you're done with all your multiplying.

Just like the commutative property, the associative property does not apply to division and subtraction. The order isn't the only thing that matters for division and subtraction; the grouping matters, too!

Take an equation that's exactly the same on both sides, such as $10 \div 10 \div 2 = 10 \div 10 \div 2$. What if you add some parentheses? You can't do that the same way you would with multiplication, because $(10 \div 10) \div 2 \neq 10 \div (10 \div 2)$; $.5 \neq 2$.

Notice that when you add the parentheses, the equation is no longer true. Why? On the left side, you divided by 10, and then you divided by 2. So both the 10 and the 2 were helping to make the 100 smaller. On the right side, you divided 10 by 2 first, which means you made the divisor a lot smaller. Now the 2 is no longer helping to make the 100 smaller, so the answer comes out differently.

There's also no associative property of subtraction. For example, look at an equation that's exactly the same on both sides: $100-40-20=100-40-20$. What if you add some parentheses in different places on each side of the equation? It doesn't work. $(100-40)-20 \neq 100-(40-20)$, because $40 \neq 80$.

Notice that when you add parentheses, the equation is no longer true. Why? On the left side, you subtracted 40, and then you subtracted another 20. So both the 40 and the 20 were helping to make the 100 smaller. But on the right side, you subtracted 20 from 40 first, so that you only end up subtracting 20 from 100. Now the 20 isn't helping to make the 100 smaller anymore, so the answer comes out differently.

So multiplication and addition have an associative property, but subtraction and division do not.

Identity Property

The identity property of a number tells you that a number always preserves its identity, or maintains its same value, when it is manipulated. The **identity property of addition** tells you that you can add zero to any number, and that number's value will remain unchanged. Adding zero to a number does nothing to change its identity. It doesn't matter what number you start with; when you add zero to a number, the value of that number will not change.

You also have an **identity property of subtraction** to consider. You can subtract zero from any number you want, and the value of that number will stay the same. While you may not have known this was called the identity property, you probably already knew that when you add or subtract nothing from a number, that action has no effect on the number.

You can remember that these are *identity* properties because they don't change the *identity* of the number. Your identity is who you are: it doesn't change depending on who's with you or what happens to you. These identity properties define certain operations you can do with certain numbers that don't cause any change to the identity of the number involved.

While the identity properties of addition and subtraction have to do with adding or subtracting zero, the identity properties of multiplication and division have to do with multiplying or dividing by 1.

The identity property of multiplication tells you that any number multiplied by 1 will maintain its same value. Multiplying a number by 1 will have no effect on its value. **The identity property of division** tells you that any number divided by one will maintain its same value. Dividing a number by 1 will have no effect on its value. This shouldn't be any surprise to you either—you already knew that nothing happens to a number when you multiply it by 1 or divide it by 1.

The identity property has a lot of overlap with **the one property** and **the zero property**. The one property tells you that any number multiplied by or divided by one will remain unchanged in value. The zero property tells you that adding zero to or subtracting zero from any number will not alter its value. Additionally, it tells you that multiplying any number by zero will result in a value of zero.

Distributive Property

The distributive property is probably not something you've had to use very much before pre-algebra, so you probably haven't thought about it in terms of numbers. Take a look at the expression $2\times(3+5)$. Now normally, you would simplify this expression by following the order of operations: you'd add $3+5=8$, and then you'd multiply $2\times8=16$. That would be true; the expression does equal 16. But notice something else that you could do: Instead of adding $3+5=8$ first, you could take that 2 and distribute it. In other words, you could think of this expression as $(2\times3)+(2\times5)$. Check and make sure you didn't change the value: $(2\times3)+(2\times5)=(6)+(10)=16$. Okay, it still equals 16. But why?

Pretend this expression is about blocks. You have three blocks plus five blocks, and you want to multiply that number of blocks by two. It doesn't matter whether you add together three blocks and five blocks (to get eight) and then double that amount, or whether you double the three and double the five and then add those values together. Either way, you're adding up all your blocks and then multiplying them by two.

Now you see why you don't really encounter the distributive property with numbers very often: you can avoid all this trouble by just following the order of operations (PEMDAS). Since you can't always do that with variables, you have to understand the rule with numbers first. Take a look at some examples just to make sure you've got the rule down pat.

The distributive property tells you that $3\times(5+2)=(3\times5)+(3\times2)$. You know that's true: $3\times7=15+6$ because $21=21$. Can you see why it's called the distributive property? You're taking the number that's in front of the parentheses and distributing it, giving it out evenly to each number in the parentheses.

You can remember that this is called the *distributive* property because it has to do with *distributing* a number. When you distribute something, you give it out evenly. So this property enables you to take something in front of the parentheses and give it out evenly to numbers that are being added or subtracted inside the parentheses.

As you transition into algebra, it will be time to start writing the distributive property in the way you're more likely to see it once variables are involved. While it's perfectly okay and completely true to write $3\times(5+2)=(3\times5)+(3\times2)$, you are more likely to see $3(5+2)=(3)(5)+(3)(2)$. These two equations mean exactly the same thing. In the second one, you've just gotten rid of the multiplication signs and used parentheses to show multiplication. That's a very common way to write problems that involve the distributive property.

The distributive property applies to subtraction in the same way it applies to addition. According to the distributive property, it must be true that $8(3-1)=(8)(3)-(8)(1)$. See what happened? The 8 that is out in front of the parentheses is distributed and then multiplied by both the 3 and the 1 individually, without changing the value of the equation. Go ahead and check: using PEMDAS, you get $8(3-1)=8(2)=16$. Using the distributive property instead, you get $(8)(3)-(8)(1)=24-8=16$.

The **distributive property** tells you that adding or subtracting two numbers and then multiplying the sum or difference by a third value yields the same result as multiplying that third value by each of the numbers and then adding or subtracting the products.

Chapter 11 Exercises

Simplify the following expressions by following the order of operations. Your answer should be in the form of one number.

1. $(2+3-1)^2-(2\times 2)$
2. $\sqrt{4}-18\div 6-3+12\div 2$
3. $\frac{9}{2}-(2+3)^2+3(2.5)$
4. $40+3.5-5^{(2-1)}\div 5$
5. $(45\div 2\times 4)\div 9+3^2$

Indicate whether each of the following expressions or equations is an example of the associative property, the commutative property, the identity property, or the distributive property.

1. $2+3=3+2$
2. $\frac{8}{1}=8$
3. $(4+9)+1=(4+9+1)$

4. $7+0=7$
5. $-5+4=4+(-5)$
6. $(2+3)+1=2+(3+1)$
7. $(14)(1)=14$
8. $3(4\times5)=(3\times4)\times5$
9. $9\times10=10\times9$
10. $3(4+1)=(3\times4)+(3\times1)$
11. $10-0=10$
12. $(13)(3)=(3)(13)$
13. $(2\cdot3)+(2\cdot5)=2(3+5)$

CHAPTER 12

Introduction to Algebraic Expressions

Once you start using variables, you won't be able to rely on your gut anymore when it comes to solving math problems. You'll really have to know what you're doing. Keep in mind that sometimes people have trouble with algebra not because of the algebra itself, but because their math skills are shaky. Now that you've got a solid foundation in the rules that govern math, you can start bringing variables into the mix and transition from arithmetic to algebra. This is where you really start doing algebra.

What's a Variable?

To start with, a variable is a letter. The variables that are most commonly used in algebra are x, y, and z, and they are the ones you'll use most often in this book. You can use any letter you want as a variable, but try to stay away from using I and O. One reason is that i has a mathematical meaning that you'll encounter if you continue studying algebra. Another reason is that I and O look a lot like 1 and 0 when you write them down, and that can lead to a lot of confusion.

So what do variables have to do with math? You use a **variable** as a placeholder for a number that you don't yet know. That's it. All of the algebra you're going to do for the rest of this book uses variables to stand in for numbers that you don't know the value of quite yet.

At this point, it's actually easiest for you to start with a word problem as a way of understanding what variables are and what they do. Here's a simple word problem: "Mary has three oranges. Joe has some oranges, too. Together, Mary and Joe have five oranges. How many oranges does Joe have?"

The problem doesn't actually tell you how many oranges Joe has. The problem tells you that Mary has three, that Joe has a mystery amount, and that Joe and Mary together have five.

When you think about the problem in your head, you might think something like, "Mary has three oranges, and when you add Joe's oranges, you end up with five in total. So three plus *what* equals five? Two!" Joe must have two oranges.

That's algebra. When you think about the problem, you're basically leaving a blank for a number you don't know. In your head, the math looks something like this:

$$3 + \_\_ = 5$$

And you know from experience that 2 is what goes in that blank, so that's the answer to the question.

If you were able to figure out that the answer to the previous example was 2, you are able to do algebra. What makes algebra tricky is that it uses a new code: a new way of writing something down that you didn't have before. And when you learn a new piece of code, you have to practice it until it becomes familiar to you. How many addition problems do you think you did in elementary school? Thousands? You did a lot of them. You've added numbers so many times that now it seems obvious and easy, and you can do it in your head. That's because you've been practicing it for *years*. Don't expect algebra to come to you instantly: it takes practice. But if you figured out that Joe had two oranges, you already have all the skills you need to do algebra. Now you just have to learn how to write it down.

ALERT

In this book, and probably in your textbook, variables are written in italics. This is usually done so that they aren't confused with regular letters when written.

All a variable does is give you a way to write the problem down without a blank, since a blank isn't really an appropriate math symbol. Instead, you use a variable. The variable is the "what" of the problem. It's the thing you need to find. So instead of writing $3+\_\_=5$, you could write $3+x=5$, using x to stand for the number of oranges Joe has. It might be easier to think of the variable as a question mark. The number you don't know is how many oranges Joe has, so that's your question mark. So when you think "$3+what=5$," it's like thinking "$3+?=5$." All a variable does is stand in for the number that you don't know.

How Do Variables Work?

Up until you study algebra, all the "blanks" in your math problems have been isolated to one side of the equal sign. For example, you might have been asked questions like "What is $3+2$?" or "$10-5=\_\_\_$." In both of these situations, you've really been doing algebra. There's something missing in the problem: some unknown for which you have to solve. In the first problem, it's expressed by the word *what*. In the second problem, it's expressed with a blank.

Using Variables

Take a look at how you would rewrite these same examples using a variable. You could write:

$$x = 3 + 2$$

$$10 - 5 = x$$

If you were to use this phrasing, you would be asked to **solve** for x. Solving for a variable means figuring out what number that variable stands for. In both of these examples, solving for x is relatively easy. In the first example, you know that x is $3+2$. That means x is equal to, or has the same value as, $3+2$. And since you know that $3+2$ is 5, you know that $x=5$. The x is actually equal to 5 in the second example too, because the equal sign tells you that x is equal to $10-5$, which is 5.

Because variables stand for numbers, they react the same way numbers react when you manipulate them. When you multiply them by zero, the product is zero. When you divide them by 1, they keep their original value. When you add 5 to them, they get 5 bigger than they were before. When you divide them by 2, they get cut down to half their size. They stand for numbers, so they react like numbers.

You can do anything to a variable that you can do to a number. You can multiply it by something or divide it by something. You can add something to it or subtract something from it. You can use it as a base or as an exponent. You can use it as the numerator or the denominator of a fraction. Anywhere you can put a number, you can put a variable.

Defining Variables

There's one big difference between variables and numbers. Any variable can stand for any number. In one problem, x might end up being equal to 10. In another problem, x could stand for 50. It could stand for –4 in still another problem, and 145 in another one. If an equation has two variables, they can each stand for any number. So in a problem that has both x and y, it's possible that x could be bigger than y. It's possible that x could be smaller than y. It's even possible that x and y could have the same value.

So any variable can hold the place of any number. However, within one expression or equation, a specific letter that is used as a variable can represent only one value. That's one of the reasons you use a variable instead of a blank or a question mark. If you wrote down $1+\_\_+\_\_=5$, you wouldn't have any idea what goes in those blanks. Maybe the blanks stand for 0 and 4, or maybe they stand for 1 and 3, or maybe they stand for .5 and 3.5. There's really no way to know.

If you wrote down $1+x+x=5$, that's a different story. Now none of those options will work, because x can't stand for 4 and 0 in the same problem. In one equation or expression or word problem, any one variable can only represent one number. So when the question had blanks, the two blanks could have been different values. But when the question has a variable, you can figure out what value that variable stands for! You are able to do this because you know those blanks are both the same number, and you know by looking at the blanks that they have to add up to 4, so the x must stand for the number 2.

If you want your variables to stand for numbers that might be different, you have to use different variables. If you don't know whether those two blanks are the same number, you'd have to write $1+x+y=5$. And now you know exactly what you knew back when you were writing blanks into the problem. Those two missing numbers must add up to 4. This is all you know. You don't know what numbers the variables represent. In other words, you don't know what either of the variables equals.

A variable can stand for any number, and it can stand for a different number in each problem where it shows up. But it can't stand for two different values in one problem. If you have more than one unknown value, you're going to have to use more than one variable.

Expressions with Variables: Simplify, Don't Solve!

An **expression** is a group of numbers, variables, and operation symbols that represents a value. You have seen a lot of expressions in your math studies so far. Here are some expressions:

$$5+2$$
$$3(20)-9$$
$$\frac{1}{4}+\frac{1}{4}$$
$$(12+4)(8-9)$$
$$7$$
$$\sqrt{4}$$
$$13^2$$

Notice something about these expressions? They don't have an equal sign. They have one or more numbers combined with one or more operators or symbols, such as + or ÷. You don't actually *solve* expressions; you *simplify* them.

There's nothing to solve for in any of these expressions. There's no mystery. You have to do the work to figure out another way to say $5+2$, but that's it. There's no unknown in the problem. You simplify these problems by completing the operations and writing the resulting number in a simpler way. For example, $3(20)-9$ has exactly the same value as 51, so you can just represent it as 51. The expression has the same value as 51. When you write the expression as 51, you're just writing it in the simplest way possible.

Simplifying expressions that don't contain any variables is something you're pretty good at by now. But what happens when you add variables? Well, the first and most important thing to remember is that you still can't solve an expression. **You cannot solve for a variable in an expression.** Take a look at a simple expression, such as $2x$. This expression has some meaning. It tells you that whatever value x has, $2x$ has twice that value. But since you don't know what x is, you cannot possibly know what $2x$ is.

Still, just like with expressions that contain no variables, you *can* simplify. For example, consider the expression $2x+4x$. Can you solve it and figure out what x is equal to? In other words, can you figure out what number the variable is standing for? No. It's an expression. You don't know what it equals, so there's no way you can find what x equals. You *can* simplify this expression, though, by writing it in a more concise or clear way that removes as many of the operation symbols as possible and leaves exactly the same value.

Combining Like Terms

The concept of combining like terms isn't really new to you. You already know that to add or subtract two things, those things have to be represented in the same units. That's why when you add or subtract fractions, you look for a common denominator. It's also why, when you do a word problem that gives some measurements in inches and others in feet, or a problem that gives some measurements of time in minutes and others in seconds, the first thing you do is put all the numbers in terms of the same unit. You can't add apples and oranges, or inches and feet, or minutes and seconds, and thirds and fifths.

That concept is the same where variables are concerned. Take the expression $2x+4x$, for example. Now, you don't know the value of x, and you can't find it, but you do know the first term in this expression is 2 times x, which means that whatever x is, $2x$ is twice as big. The second term, $4x$, is four times as big as x. So you have two xs and four xs, and you're adding them together. You must have six xs all together. That's because x is the unit these terms have in common, and each term is telling you how many units you have. Because x is common to both terms, you can combine these terms and rewrite them as $6x$. This is called *combining like terms*, or taking terms that are in the same units and combining them.

If it helps, you can think of x as a tangible thing for the moment (even though it's not a thing; it's a number). The expression above could be read as "two exes plus four exes," which is the same as saying "two oranges plus four oranges."

When you're looking for like terms, look for the same variable raised to the same power. Two terms with no variables are always like terms. Two terms with the same variable are like terms. Two terms with the same variable squared are also like terms.

The "oranges" part stays the same, because that's the unit that you're adding. When you add two oranges and four oranges, you're really only adding 2 and 4. The word *oranges* just tells you what these numbers are counting. The original expression $2x + 4x$ is the same; nothing happens to the x itself. When you add two of something to four of something, you get a total of six of that thing. In our case, the "thing" just happens to be the unknown quantity that x stands for.

Counting the Terms in an Expression

What exactly are the terms of an expression? What does the word *term* mean in this context? **Terms** are the pieces of an expression that are combined by addition or subtraction. The minimum number of terms that an expression can have is one. An expression that has only one term is called a **monomial**. A single number, such as 100 or 4, is a monomial. So is a single variable, such as x or y. Other monomials are messier—as long as the numbers and variables are combined by something other than addition or subtraction, they only count as one term. So, xyz is a monomial. So is $2x$. So are y^2 and $\frac{x}{8}$ and $\sqrt{4x}$ and $(3 \div 3)$. Remember, you don't decide how many terms you have based on how many letters and numbers are involved or how messy the expression looks; you decide based on how many parts of the expression are separated by a + or a –.

Watch out for plus or minus signs that are enclosed in parentheses. Because of the order of operations (PEMDAS), you know that parentheses link whatever is inside them together. For that reason, if you have addition or subtraction inside a set of parentheses, you get to treat it as one term. Because of this rule, an expression such as $2(3 + x)$ is still a monomial expression: there is only one term involved.

Remember that parentheses come first, regardless of what operation is inside them. Also, when an operation, or set of operations, is in the numerator or denominator of a fraction, you can treat it as if it's in parentheses.

More than One Term

When an expression has two terms, you call that expression a **binomial**. The prefix *bi–* often means "two," which is why a bicycle has two wheels and binary code uses only two types of numbers. A binomial is one type of **polynomial**, which is an expression that has more than one term. So if an expression has one term, it's a monomial. If it has more than one term, it's a polynomial. If it has exactly two terms, it's a special type of polynomial called a binomial. These words give you a shared way to talk about math clearly and to write down the rules you want to follow. The terms themselves may also be on your class tests, but most important, you want to understand what your teacher or textbook is talking about when these words are used.

The term *polynomial* is a term you're going to use and hear throughout the rest of your math studies. The *poly* part means "many." For example, a polygon has many sides, a polyglot is a person who speaks many languages, and a polygraph measures a variety of bodily changes (to see if a person is lying or not).

What does a polynomial look like? It can show up in many different forms, but as long as there are some units being combined by addition or subtraction that is not in parentheses, you've got a polynomial. So, for example, $6+3$ is a polynomial. So is $x+y$. So is $7-y$. Your polynomial might be messier and look like $\frac{2}{7}-9x$ or $x^2+34-6y$. These are all polynomials, which simply means that they have more than one term. $6+3$, $x+y$, $7-y$, and $\frac{2}{7}-9x$ are all polynomials of the special type known as binomials: they have exactly two terms, because they have two groups of numbers, symbols, and operators that are joined with a plus or minus sign. $x^2+34-6y$ has three terms. There's no limit on the amount of terms an expression can have! For example, $x+y+2-3+4x+y-2$ has seven terms!

Chapter 12 Exercises

Indicate how many terms there are in each of the following expressions.

1. $4x+9x$
2. $3y+9+16y-3$
3. $8xyz$
4. $8(12+4)$
5. $3y-2(x+9+3y)$
6. $a-b+3c$
7. $\frac{4x}{x+y}$
8. $4x+x(3-2x)$
9. $4a+6b$
10. $10(x-x-x)$

Each of the following expressions has two like terms. Identify the two like terms.

1. $4x+9+3x$
2. $2-3x+9$
3. $2y+4y+6$
4. x^2+2x^2+y
5. $x^2-2x-2x^2$
6. $x+y+z-x$
7. $\frac{3+4}{2}+2x+9$

CHAPTER 13

Manipulating Expressions with Variables

By its very definition, algebra is doing math that uses variables. In pre-algebra, you spend time building the skills you need in order to understand and work with these variables, and you also start doing algebra. In this chapter, you will learn how to start manipulating expressions with variables. That means taking one way of writing something with a variable, and writing it in a different way that doesn't change the value being expressed.

Simplifying Expressions with Addition and Subtraction

When you're asked to simplify an expression that includes addition and subtraction, you have to start by realizing that you can only simplify like terms. Here are some expressions you *cannot* simplify:

$$8+x$$

$$x+y$$

$$2x+10+y$$

Why can't they be simplified? Look at the variables in the terms of each expression. In the first expression, you have a term with no variables and a term with an x. Those aren't like terms (meaning they don't have the same variable or product of variables), so they cannot be added or subtracted. The second expression has one term with an x and one term with a y; these are not like terms, so they can't be combined. And none of the terms in the third expression are alike, so there's nothing you can combine.

If the terms are alike, you can simplify the expression. What kind of terms count as like terms? Terms that you can combine by adding and subtracting. The easiest like terms to think about are numbers that don't have any variables with them. Numbers by themselves can always be combined by addition or subtraction with any other numbers that are by themselves. So you can simplify $3+5$ and rewrite it as 8. You can simplify $\frac{2}{3}-6$ or $8+\frac{14}{3}$ and write each of these polynomials as a monomial. In other words, any time two or more of your terms are numbers by themselves, you can combine them to simplify the expression. Square roots, although they are numbers, can only be added or subtracted if they can be represented as the same root of the same number. For example, $2+\sqrt{2}$ and $\sqrt{3}-\sqrt{2}$ are examples of binomials that cannot be further simplified, but $\sqrt{2}+\sqrt{2}$ *can* be further simplified. In the previous expression you are adding a one "$\sqrt{2}$" to another "$\sqrt{2}$", which gives you two of them, or $2\sqrt{2}$.

You can also combine any terms that share the same variable or combination of variables. In the example you saw earlier, you simplified $2x+4x$ to $6x$. You can do this because the two terms are alike: they both tell you how many of x you have. So when two terms each have the same variable, they are like terms, and you can combine them by adding or subtracting to simplify the expression to only one term. For example, you can simplify $5y-3y$ and rewrite it as $2y$. Think of y as a tangible thing: you have five of them, and then you take three away, so you only have two left.

When it comes to combining like terms, roots are a lot like variables. For two terms to be "alike," and for you to be able to combine them by addition or subtraction, they must contain the same variables (if any) raised to the same power, and they must contain the same roots (if any).

Simplifying Expressions with Multiplication and Division

In order to simplify an expression that includes addition or subtraction, you need to have like terms. Just as you can't add apples and oranges, you can't add x and y. Well, you can add them, but you can't write their sum in any simpler form than $x+y$. However, you do have a few ways of writing multiplication and division that might be new to you.

Writing Terms Out

Say, for example, you're asked to simplify $8 \div 3$. You already know that another way you can write this same expression is $\frac{8}{3}$, because a fraction bar is one way to show division. What if you're asked to simplify $x \div y$? You can follow exactly the same rule and rewrite that expression as $\frac{x}{y}$. You haven't solved anything: you still don't know what the variables equal. You also haven't changed the number of terms: both expressions have only one term. But you've written the expression as one number instead of two, and therefore, you have written it in a way that's easier to use.

Now let's take a look at the expression $2x \div x$. First, take note of the fact that if you want to, you can rewrite this expression as $2x \div 1x$. You can always multiply any variable you want by the number 1, because multiplying by 1 doesn't change the value of a number.

So what about this expression—can you simplify it? Sure. You can rewrite it, just as you can rewrite any division problem, and change it to look like this: $\frac{2x}{x}$. You haven't changed the value of the expression at all. You've just written it in a way that's easier for you to use and understand, which is all simplifying is. But this time, you're not done! Take a look at the fraction $\frac{2x}{x}$. Notice that the numerator says $2x$, which is just another way of writing $2 \times x$. (See why you stay away from the $\times$ sign for multiplication when you start dealing with algebra? It looks so much like the x that it is very confusing!) So now you have a fraction that has x as a factor in the numerator and the denominator, which means that you can factor the x out of both. Just like you did with fractions that did not have variables, you can divide the numerator and denominator by the same value and not change the value of the fraction. In other words, both the top and bottom of this fraction have x as a factor, and you can take it out of both. What will you have left when you do? You'll have $\frac{2}{1}$, which simplifies to 2. That means that this expression must always equal 2, no matter what the variable x stands for.

Skeptical? Try it. What if x stands for 7? Then $2x \div x$ would be the same as $2(7) \div 7$, which equals 2. What if x stands for 3? Then $2x \div x$ would be the same as $2(3) \div 3$, which equals 2. See what's happening? It doesn't matter what number x stands for. Whatever number you plug in, you're going to multiply 2 by that number and then immediately divide 2 by that number. So you'll always end up with 2 as the answer, no matter what value you assign to x.

Just Remember

When you first start learning to simplify expressions, it can seem overwhelming. Remember that variables stand for numbers, so they can do everything numbers can do, and they can't do anything numbers can't do. So if you could rewrite a division problem as a fraction when it was made out of numbers, you can rewrite a division problem as a fraction when it's made out of variables. Similarly, if you can rewrite 7×2 without the multiplication sign by writing the numbers next to each other in parentheses like $(7)(2)$, you can do the same thing with variables. So, if you wanted, you could rewrite $y \times z$ as $(y)(z)$. In fact, you'd probably want to do that so it wouldn't be so confusing to read! And remember, with variables you don't even need the parentheses: you can rewrite $(y)(z)$ as yz, and your expression would have the exact same value as when it was written the original way.

Simplifying Expressions by Adding and Subtracting Fractions

You already know the rules for working with fractions, and those rules don't change when variables are involved. That's why you took the time to practice the rules—because you need to be comfortable with them before you throw variables into the mix.

For example, from working with numbers, you know the rules of adding and subtracting fractions: find a common denominator and then add or subtract the numerators. So what if you're asked to simplify the expression $\frac{x}{3}+\frac{5}{3}$? Well, you see fractions with addition, so you know you need a common denominator. Since you already have one, all you have to do is keep the denominator and add the numerators together. So you can rewrite $\frac{x}{3}+\frac{5}{3}$ as $\frac{x+5}{3}$. After this, there's not much else you can do, because x and 5 are not common terms, so they can't be combined by addition.

No matter what value x stands for, this rephrasing will still be true. Try plugging in any number you want for x, and you'll see that no matter what number you pick, $\frac{x}{3}+\frac{5}{3}=\frac{x+5}{3}$. That's what simplifying expressions does. It doesn't change their value; it just creates a new way to express the same value.

What if the denominators aren't the same? You'd have to make them the same, just like you did when the fractions didn't have variables in them. For example, what if you're asked to simplify $\frac{x}{5}-\frac{3}{10}$? Well, you need the denominators to be the same, so you're going to make them both 10, because that's the least common multiple of 5 and 10. That means you need to multiply the first denominator by 2, and since you don't want to change the value of the fraction, you're going to need to multiply the numerator by 2 as well. In other words, you're going to multiply the first fraction by $\frac{2}{2}$ and rewrite the expression as $\frac{2}{2} \bullet \frac{x}{5} - \frac{3}{10}$. Now you've got to do a little multiplication, but that's no problem, because you know the rules for multiplying fractions: multiply the numerators and multiply the denominators. So now you've got $\frac{2x}{10} - \frac{3}{10}$. At this point, you can follow the normal rules of adding and subtracting fractions: keep the denominator and subtract the numerators, leaving you with $\frac{2x-3}{10}$. You can't simplify any further, so you're done!

When in doubt, go back to the rules of fractions and let them tell you what to do. For example, what if you want to simplify $\frac{3}{x} + \frac{2}{x}$? The rules of fractions tell you to keep the common denominator and add the numerators, so you can rewrite this expression as $\frac{5}{x}$. The rules don't change just because there are variables involved.

Simplifying Expressions by Multiplying and Dividing Fractions

Just as with adding and subtracting, the fraction rules that you learned with numbers will work with variables. If you're asked to simplify $\frac{3}{x} \times \frac{x}{2}$, you should follow the rules for multiplying fractions: multiply the numerators to get the new numerator, and multiply the denominators to get the new denominator. When you do that, you have $\frac{3x}{2x}$.

Even though the bottom is $x \bullet 2$, when you combine a variable and a number in a multiplication expression, you always write the number first. So even though $x2$ and $2x$ technically mean the same thing and have the same value, the code is always written as $2x$.

Now, just as you would with numbers, you can follow the rules of fractions and take out any factor that's shared by the numerator and the denominator. A factor of a number is anything you multiply by to get that number, so both $3x$ and $2x$ have x as a factor. Therefore, you want to divide both the numerator and the denominator by x, because doing so will simplify the expression without changing its value. So now you have $\frac{3x \div x}{2x \div x}$, and with those xs factored out, you're left with $\frac{3}{2}$. Even though you have no idea what x is, you know that the value of the expression $\frac{3}{x} \times \frac{x}{2}$ will always be $\frac{3}{2}$, because you are able to simplify it to get the value of $\frac{3}{2}$. You can test this conclusion by plugging in any number you want for x in the above expression. You'll see that no matter what value you pick, that value will cancel itself out, and you'll be left with $\frac{3}{2}$.

Simplifying Expressions with Exponents and Roots

At this point, you should be starting to see that because variables stand for numbers, you have to treat them like numbers. The same is true when you look at expressions that involve exponents and roots. All the old rules apply. Take look at the old exponent and root rules, but this time check them with variables involved.

1. **When you multiply numbers with the same base, add the exponents.** So this rule tells you that $x^2 \bullet x^5 = x^7$. It doesn't matter what number x is, you're taking it as a factor two times and then five more times, which is the same as taking it as a factor seven times.
2. **When you divide numbers with the same base, subtract the exponents.** This rule tells you that $\frac{y^9}{y^3} = y^6$. It doesn't matter what number y is, you're taking nine of something multiplied together and dividing three of those things out of the expression, which leaves you with six of those things multiplied together in the end. You can plug in any number

you can imagine for *y*, and the simplified version of the expression will have the same value as the original expression.

3. **When you multiply square roots, you can multiply the numbers under the square root and take the square root of that product.** This rule tells you that $\sqrt{x} \bullet \sqrt{x} = x$.Why? Because you can change $\sqrt{x} \bullet \sqrt{x}$ into $\sqrt{x \bullet x}$, which is the same as $\sqrt{x^2}$. When you square a number and then take the square root of your result, you're just doing something and then undoing it. The square and the square root cancel each other out. Check it out with numbers to prove it: $\sqrt{4^2} = \sqrt{16} = 4$. You can take any positive number you want—if you square it and take the square root of it, you end up right back where you started.
4. **When you raise an exponential expression to another exponent, multiply the exponents.** You learned this rule with numbers, but it's a rule, so it must hold true for variables as well. Therefore, you know that you can simplify $(x^3)^{10}$ by multiplying the exponents, which means that $(x^3)^{10} = x^{30}$.
5. **When you add and subtract with exponents, you can only count up like terms. If the terms aren't alike, you can't simplify.** This is just the same as it was for numbers. So if you're asked to simplify $x^2 + y^2$, there's nothing more you can do. That's as simple as the expression is going to get. However, if you are asked to simplify $x^3 + x^3 + x^3 + x^3$, you can realize that you're adding four of the same term, and thus your sum will be $4 \bullet x^3$ or $4x^3$.
6. **When you square a square root, you cancel it out.** If you simplify $\sqrt{4}^2$, you'll see that you end up with 4, because $\sqrt{4} = 2$ and $2^2 = 4$. When you take the root and then square that root, you end up right back where you started. This is true for any number, so it's also true for any variable. Therefore, $\sqrt{x}^2 = x$. No matter what positive value you plug in for *x*, this equation will be true.

Why are you doing all this simplifying?
Expressions are the building blocks of equations. You need to be comfortable with the rules for simplifying expressions before you start putting those expressions into equations. Once you look at equations, you'll see that equations and how they're set up will force you to manipulate expressions in order to solve for the variable.

Chapter 13 Exercises

Simplify the following expressions to the furthest extent possible.

1. $9x-2x$
2. $8y+3y$
3. $x+2y$
4. $x-4x$
5. $a+3a$
6. $b-b$
7. $b-a$
8. $x+x+1$
9. $x-x-x-x$
10. $2y+3y-2y$
11. $2y\times3$
12. $x\bullet x$
13. $\frac{b}{b}$
14. $\frac{4a}{2a}$
15. $8x\div4x$
16. $8x\div4$
17. $10\times y$
18. $x\bullet y\bullet z$
19. $z\bullet z\bullet z\bullet3\bullet2$
20. $x\div5$
21. $x+\frac{1}{2}$
22. $\frac{2}{3}+\frac{a}{3}$
23. $\frac{b}{3}-\frac{b}{6}$
24. $\frac{1}{x}+\frac{1}{x}$
25. $\frac{x}{2}+\frac{x}{2}$
26. $\frac{b}{2}-\frac{b}{4}$
27. $\frac{2}{s}+\frac{s}{2}$
28. $\frac{b}{3}-\frac{b}{3}-\frac{a}{2}$
29. $\frac{x}{y}+\frac{x}{y}$
30. $\frac{d}{d}+\frac{d}{d}+\frac{d}{d}$
31. $\frac{x}{3}\times\frac{3}{x}$
32. $\frac{4}{y}\bullet\frac{3}{y}$
33. $\frac{x}{3}\times\frac{x}{3}$
34. $\frac{4x}{2}-x$
35. $y+\frac{1}{3}y$
36. $\frac{5}{3}x+\frac{1}{3}x$
37. $\frac{3}{y}+\frac{3}{y}+\frac{3}{y}$

38. $\frac{3x}{x} \bullet \frac{x}{x}$

39. $\frac{k}{2} - \frac{3k}{2}$

40. $\frac{x}{5} \times \frac{4}{x}$

41. $x \bullet x \bullet x$

42. $y \bullet 2y$

43. $\sqrt{y}\sqrt{y}$

44. $\sqrt{d}\sqrt{d}\sqrt{d}$

45. $\frac{\sqrt{8x}}{\sqrt{2x}}$

46. $(x^2)^3$

47. $(\sqrt{b})^4$

48. $x^3 + x^3$

49. $y^2 - y^2$

50. $y \bullet y \div y$

CHAPTER 14

Introduction to Equations

Expressions, as you know, can only be simplified, or written in another form—they cannot be solved. What *can* you solve? You can solve equations. Equations are the real heart of algebra. Once you understand how equations work, there's nothing in algebra that you won't be able to do. This chapter discusses what makes an equation, what you're supposed to do with an equation, and how to do it.

What's an Equation?

An **equation** relates two quantities with an equal sign. If expressions are the words and phrases of math, equations are the sentences. Equations tell you that two quantities are equal, which means that those two quantities are the same in every way. Here's a simple equation that probably seems obvious to you: $5=5$. First of all, it has two quantities on opposite sides of an equal sign. That equal sign works like a scale that is perfectly balanced: it tells you that whatever is on the left side is the same as whatever is on the right side.

The two quantities being compared by the equation look exactly the same. They don't have to look exactly the same though. As long as both sides have the same value, the equation will be true. Think of the equal sign like the balance of an old-fashioned scale. As long as the weights on both sides of the equal sign balance, the equation will be true. So, for example, here's another true equation: $2+3=5$. You know that this is true just by looking at it because you are used to these numbers. What does it really mean? It means that when you add together two and three, the result will have the same weight as the number five on the other side of the scale. That's what makes these two quantities equal.

The reason you haven't seen equations before preparing to study algebra isn't that you can't write them with numbers; it's that there's really no point to writing an equation with just numbers. Up until now in your math studies, you've mostly been practicing arithmetic. If you set up an arithmetic problem that says $2+3=5$, there's nothing for you to do!

What's the Point? Solve for the Variable!

When you're given an equation, the goal is to solve for the variable. Solving for the variable means finding what it equals. In other words, you've solved for the variable when you know what number it stands for. Consider a simple equation such as $x=9$. This is an equation, because it has an equal sign equating two values. You know that the value on the left side of the equation has to be equal to the value on the right side of the equation. Therefore, the value of x is 9.

That was easy. So when there's a single variable on one side of the equation and a single number on the other side, solving isn't any work at all. Now take a look at an equation such as $x=\frac{(3+2)}{9}$. Well, it's not as easy as it was before, but it's clear what you have to do. You already have x all by itself, but the other side of the equation has an expression that needs to be simplified. So in order to find the value of x, you need to simplify the other side of the equation: $\frac{(3+2)}{9}=\frac{5}{9}$. Since those values are exactly the same, you can replace one with the other in the equation. Therefore, you know that $x=\frac{5}{9}$. Neither expression can be further simplified, so you're finished. You've solved for the variable x, which means you know that it represents the value $\frac{5}{9}$. This is the one and only value that x can stand for in this problem because it's the only value you can plug in for x that keeps the equation true.

Remember, every variable is really just a blank. Whatever you put into that blank has to make both sides of the equation equal to each other. Your goal with any equation is to figure out the value of the variable or variables in that equation. And you do this by a process called isolating the variable.

Isolating the Variable

To **isolate the variable** means to get the variable alone on one side of the equation. If you're isolated from your friends, you are away from them. If the variable is isolated, it's away from everything else in the equation. Once the variable is by itself, you're able to solve the equation. If you have the variable on one side of the equation and all your numbers on the other side of the equation, it doesn't matter how many numbers there are or how they are arranged. You will be able to simplify them as much as possible, and that will tell you the value of the variable. For example, here's a big, messy equation: $x=(2+4)(3-9)+\frac{1}{(6\times3)}+(4^2-6)$.

Okay, so maybe you don't *want* to solve that equation for x, but you could. The variable is isolated, so you know that its value is the same as the value of whatever's on the other side of the equation. You can follow your PEMDAS rules and simplify the other side of the equation to find out that $x=-25\frac{8}{9}$. That is the only value you can plug into the "blank" of x that is equal to the big mess on the other side of the equation.

When you see "plug in," it means "replace." It helps to think of each variable as a blank: just like you'd "fill in" a blank, you can "plug in" a number for a variable by putting that number where the variable is.

But what do you do if the variable isn't already by itself? How do you isolate it? Good news: Variables may be pretty new to you, but they follow all the same rules as arithmetic with numbers. Equations may be very new to you, but there is really only one new rule you have to learn to work with them. It's the most important thing to know about equations, and it's true about every equation you'll ever encounter in math for the rest of time. Ready? Here's the rule: **As long as you do the same thing to both sides of an equation, the equation will still hold true.**

To isolate the variable, you need to undo whatever's being done to it. Think of it this way: if you had two people around you and you wanted to isolate yourself, you'd have to get rid of those people.

What does this rule mean? Consider an equation you know and understand, like $5=5$. This equation is true, because the value on the left is the same as the value on the right. That's what makes an equation true. Picture an equation again as a scale on which the quantity on the left exactly balances the quantity on the right. Just as long as you do the same thing to both sides, the scale will still be balanced. The numbers on the right and left side might not be the same as they were, but they will still be equal.

What kind of things can you do to both sides?

- **Add:** You can add any value you want to both sides of an equation, and the equation will still be true. If you take your $5 = 5$ equation, a perfectly balanced scale, and add 2 to both sides, what happens? Now you have $5 + 2 = 5 + 2$. That's still true! The value on the left and the value on the right are both 7, so the equation is still true.
- **Subtract:** You can subtract any value you want from both sides of an equation, and the equation will still be true. If you take your $5 = 5$ equation, a perfectly balanced scale, and subtract 3 from both sides, what happens? You have $5 - 3 = 5 - 3$. That's still true! The value on the left and the value on the right are both 2, so the equation is still true.

Notice that the actual values are changing, but it doesn't matter. The only purpose of an equation is to tell you that two values are equal. As long as you change both values in the same way, the equation will still be a true equation. You don't care that it's telling you the true information in a different way; you only care that it's still giving you true information.

- **Multiply:** Return to $5 = 5$, a nice balanced equation. What happens if you multiply both sides by 30? Well, your equation changes into $150 = 150$, but that's still a true equation! You can multiply both sides by any value you want, and the truth of the equation will remain.
- **Divide:** You can take $5 = 5$ and divide both sides by 5, and you'll be left with the true fact of $1 = 1$.

Beyond the Basics

Adding, subtracting, multiplying, and dividing are the most common operations for you to do in math, but they aren't the only things you're allowed to do to both sides of an equation. You can square both sides. You can take the square root of both sides. You can add a value to both sides, and then multiply both sides by another value. You can do anything you want, and as long as you do the same thing to both sides of the equation, you'll get another true equation.

Solving Equations with Addition and Subtraction

Say you're asked to "solve for x" (which is a pretty common set of directions for an algebra problem) in the equation $x+2=7$. You might just be able to think about this equation in real-world terms and find the answer. This equation is basically telling you that some number, when you add 2 to it, will give you 7. With a little thinking and some trial and error, you can figure out that the number you are looking for is 5. It is the only number that can fit in the blank, because 5 is the only number that turns into 7 when you add 2 to it. So you've solved the equation: you know that $x=5$. You can plug 5 into the equation for x and get a true statement, because $5+2=7$.

Solving More Difficult Equations

What if you didn't know that the answer was $x=5$? What if you couldn't figure it out in your head? What if the numbers were bigger or messier? That's where the rules of algebra come along to save the day. Remember, when you want to solve for the variable in an equation, you need to isolate that variable, or get it all by itself on one side of the equal sign. Take a look at the equation $x+2=7$. The x isn't by itself, and it needs to be. Right now, the x has a $+2$ attached to it, and you need that to be gone. So how can you get rid of it? What could you do to that side of the equation to cancel out the $+2$ and make it go away? In other words, what operation could you perform to undo $+2$ and give it a value of zero?

If you haven't guessed, the answer is you could subtract 2. That would solve your problem, because it would get rid of the 2 that is currently being added to the x. But can you just subtract whatever you want? It seems like that wouldn't be allowed. This is where the rule you just learned comes in: you can do anything you want to an equation *as long as you do the same thing to both sides*. So you *can* subtract 2 from the left side of the equation, just as long as you also subtract it from the right side of the equation.

Check it out: you start with $x+2=7$. Then, to isolate the variable, you decide you need to subtract 2 from the left side. Because this is an equation, you need to do the same thing to both sides, so you're going to subtract 2 from both sides. Now your equation looks like this: $x+2-2=7-2$. So far it looks worse instead of better! But you're not done yet. Your final step is to combine like terms in order to simplify the equation. On the left, you have two numbers without any variables, so you can combine those. Adding 2 and then subtracting 2 is the same as adding zero, so on the left side of the equation, you know that $x+2-2$ has the same value as x. On the right side of the equation, your simplification is really easy: $7-2=5$. So, now you know that $x=5$.

If an amount is being added to your variable and you want to get rid of it, subtract that amount from both sides of the equation. And if you undo addition by using subtraction, how do you think you undo subtraction? You use addition.

Take a look at the equation $y-8=6$. You need to solve for y, which means you need to isolate y. Right now there is a –8 on the y side of the equation that you need to get rid of to get the y by itself. How do you cancel out a minus eight? You add eight to it. Since this is an equation, if you add 8 to one side, you have to add it to the other side, too! So in order to isolate the variable, you add 8 to both sides. Now you have $y-8+8=6+8$. The left side of the equation simplifies to y, and the right side of the equation simplifies to 14. Now your new, true equation is $y=14$. In other words, 14 is the number that y is standing for in the equation.

You can double-check your work by replacing y with 14 in the original equation. Since you're saying that y and 14 are equal, the equation should still be true if you replace one with the other. Let's see. If you replace y with 14 in the original equation, you'll have $14-8=6$, which is true! So you know you've done it right.

Solving Equations with Multiplication and Division

In the last section, you saw that addition and subtraction are opposites: you can use one of them to undo the other. If you add a value and then subtract the same value, you are right back where you started. So when you have a value being added that you don't want, you can get rid of it by subtracting that value. Likewise, when you have a value being subtracted that you don't want, you can get rid of it by adding that value.

Guess what? Multiplication and division are opposites, too: you can use one of them to undo the other. If you multiply by a value and then divide by the same value, you come right back to where you started. And when you have a value being multiplied that you don't want, you can get rid of it by dividing by that value, and when you have a value being divided that you don't want, you can get rid of it by multiplying by that value.

Examples of Multiplication and Division with Variables

What if you're asked to solve the equation $2x = 8$ for x? No problem! You need to solve for x, which means you have to get x by itself. But now there is an obstacle: x is currently being multiplied by 2, and you need to get rid of that 2. How? If you want to undo multiplication, you must use division. So you're going to go ahead and divide both sides of the equation by 2. Dividing both sides by 2 changes the equation to $\frac{2x}{2} = \frac{8}{2}$.

Now you have to simplify. $\frac{8}{2}$ is the easier side, because you know that $\frac{8}{2}$ is equal to 4. How do you simplify the $\frac{2x}{2}$ side of the equation? The same way you would if these were all numbers instead of a mix of numbers and variables. Take a look at the numerator and the denominator and notice that they both have 2 as a factor. Because of this, you want to divide both the numerator and denominator by 2 in order to reduce the fraction without changing its value. When you do that, you'll find that $\frac{2x \div 2}{2 \div 2} = \frac{x}{1} = x$. So now you've isolated x. Your final, simplified equation tells you that $x = 4$.

You can plug that value back in for x in the original equation to double-check your work and make sure you are correct. You are, because $2\times4=8$.

For now, you should write out this division to make it very clear why the x is isolated by this action. Eventually, you'll be able to quickly spot that when you take $2x$ and divide it by 2, those 2s will cancel and only the variable will remain. It doesn't matter what the number is: if you multiply and divide a variable by the same number, it is the same as multiplying by 1, which has no effect on the value of the variable.

Not surprisingly, just like you can undo multiplication with division, you can undo division with multiplication. What if you're asked to solve for x in the equation $\frac{x}{3}=5$? Well, to solve for the variable, you need to isolate the variable. Right now, the variable is divided by 3, and you don't want that. So what do you do? You have to perform an operation that will cancel out dividing by 3. Since multiplication is the opposite of division, you know you need to multiply by 3, and since this is an equation, if you're going to multiply the left side of the equation by 3, you need to multiply the right side of the equation by 3 as well, so that both sides will still be equal. Once you multiply both sides by 3, you'll see that the equation now says $\frac{x}{3}\times3=5\times3$. The right side of the equation is clearly easier to deal with: it simplifies to 15.

What happens over on the left side? Well, you're taking whatever x is, dividing it by 3, and then multiplying it by 3. When you divide something by 3 and multiply it by 3, you come back to whatever you started with. So the left side of the equation simplifies to x. You can double-check this logic using your rules for multiplying fractions: $\frac{x}{3}\times3=\frac{x}{3}\times\frac{3}{1}=\frac{x}{1}=x$.

Now you can see why fraction rules are so important—they are essential to your ability to multiply and divide when solving equations! Remember that any time you multiply and divide by the same number, those actions cancel one another out, just like they do in a fraction.

So, your final equation is $x = 15$. Once again, you can double-check your result by plugging 15 in for x in the original equation, and you'll see that you're correct, because it's true that $\frac{15}{3} = 5$.

Solving Equations with Exponents and Roots

Pre-algebra is tricky because it involves a lot of new rules coming together at once. It's sort of like learning to drive—you can talk about one part of driving at a time and understand it, but to actually get behind the wheel and make the car go, you're going to have to do a lot of new things all at once. In driving, you can't always stop and do things one step at a time. (The people behind you on the highway probably wouldn't be very happy if you did.) But in pre-algebra, you can. So if you're still finding the rules for exponents and roots tricky, this would be a good time to break out your flashcards or notes and review the rules. You might even find it helpful to have the rules in front of you while you're reading this section.

Adding a Variable

All the rules that apply to exponents and roots when dealing with numbers apply to exponents and roots when dealing with variables. For example, if you're asked to solve for x in the equation $x = \sqrt{16}$, you know what to do. You just simplify the right side of the equation to get $x = 4$. You can even make the right side of the equation messier; for example, maybe you're asked to solve for x in the equation $x = \sqrt{8} \times \sqrt{2}$. The x is already isolated, so the algebraic part of your work is done. Now you have to do some arithmetic over on the right side of the equation. You can follow the shortcut for multiplying with square roots and rewrite the equation as $x = \sqrt{(8 \times 2)}$, and then simplify even further using the order of operations to get $x = \sqrt{16}$. Now you know that $x = 4$. So far, you're manipulating exponents and roots just as you would if the x weren't involved, because the x is already isolated on its own side of the equation.

But what happens if x is not isolated on one side? What if you're asked to solve, for example, $\sqrt{x} = 4$? You might be able to just think about this equation and figure out what x must equal. Remember that the variable works like a blank in the equation. It's the "what" that you're being asked to find. So when you're asked to solve $\sqrt{x}=4$, you're really being asked, "What number, when you take the square root of it, will give you four as the answer?" In other words, "Four is the square root of what number?" This is actually on your perfect square memorization list, so you know the answer: 4 is the square root of 16, so x must equal 16. You can double-check your thinking by plugging 16 in for x in the original equation and making sure it's true. If you do that, you'll have $\sqrt{16}=4$. Right!

So that's how you get there by just thinking about the problem in your head, but what's the algebraic rule you can follow to be able to get the answer systematically? You already learned that addition and subtraction are opposites: when you want to get rid of one of them, you use the other one. You have already learned that multiplication and division are opposites: when you want to get rid of one of them, you use the other. And now you will learn that exponents and roots are opposites: when you want to get rid of one of them, you use the other.

Go back to the equation $\sqrt{x}=4$. You know that your goal is to solve this equation, and that the way you solve an equation is by isolating the variable. You need to get the x by itself. Right now, it is under a square root sign. How do you undo a square root? By squaring it. And if you're going to square one side of the equation, you have to square the other side as well. So you take $\sqrt{x}=4$ and change it to $\sqrt{x}^2=4^2$. The right side of the equation is all numbers, so you can simplify it to 16. The left side reads $\sqrt{x}^2$, which means whatever x is, you've taken the square root of it and then squared the result. Those actions cancel one another out, so you're left with just x. Now your equation says $x=16$, and you're done. You know that 16 is the only value x can have, so you're finished solving the equation.

Undoing a Squared Variable

Just as you can undo a square root by squaring it, you can undo a square by taking its square root. If you're asked to solve the equation $x^2=64$, you know that you need to isolate the variable. Right now, it's being squared, so to undo that operation, you need to take the square root. And if you're going to take the square root of one side, you need to do it to the other side as well to keep the equation an equation. Now you've got $\sqrt{x^2}=\sqrt{64}$. Since $\sqrt{64}=8$, you can rewrite the equation as $\sqrt{x^2}=8$. It's true that the squaring and square rooting on the other side of the equation cancel each other out, but there is one extra step you have to watch out for! It turns out that you could plug in 8 or –8 for x, and this equation would still be true. Whether you plug in 8 or –8, when you square it, you'll get 64. Then, when you take the root of 64, you'll get 8. That means that in this case, there are two possible values of x: both 8 and –8. You would write the solution as $x = 8, -8$ to show that those are the two possible solutions.

FACT

The rule to remember is that $\sqrt{x^2}$ can equal x or −x. However, $\sqrt{x}^2$ can equal only x. Why? Because it's impossible to take the square root of a negative number—there's no value that can be multiplied by itself to give you a negative number. So in $\sqrt{x}^2$, x can only be positive.

Chapter 14 Exercises

Isolate the variable in the following equations. Do not combine or simplify any of the numbers or solve for the variable—just do the first step.

1. $x-2=10$
2. $3+y=-8$
3. $b+4=0$
4. $a-3=3$
5. $3a=9$
6. $\frac{x}{2}=7$
7. $8x=8$
8. $x^2=25$
9. $\sqrt{x}=4$
10. $x^2=13$

Solve the following equations.

1. $x+3=13$
2. $x-10=26$
3. $2x=10$
4. $\frac{x}{4}=9$
5. $x^2=81$
6. $\sqrt{x}=9$
7. $3+b=15$
8. $b-3=15$
9. $3b=15$
10. $\frac{b}{3}=15$
11. $b^2=1$
12. $\sqrt{x}=0$

CHAPTER 15

Number Properties (with Variables)

You already know many of the rules and properties of numbers, but now you're going to add in some variables. This chapter is going to take what you already know about numbers and show that it's also true about variables, because, as you know, variables stand for numbers, so they behave like numbers.

Commutative Property

You know that addition and multiplication have a commutative property. You also know that the commutative property says you can rearrange the order of numbers being added or multiplied without changing the result. You've seen this with numbers: $2 \times 3 = 3 \times 2$ and $2 + 3 = 3 + 2$. Because the commutative property applies to addition and multiplication with numbers, it also applies to variables. Thus variables. So $x + y = y + x$ and $xy = yx$. No matter what numbers you plug in for x and y, these equations will be true! Because multiplication and addition each yield the same result regardless of the order in which they are performed, they each have a commutative property: you can write them in whatever order you want.

A property is just a fact that's true about something. So, for example, having four legs is a property of dogs, and being orange is a property of oranges. These properties of addition and multiplication are simply things that are true about addition and multiplication.

Identity Property

There are four different identity properties. The identity (or value) of a number doesn't change when you add or subtract zero from it. No matter what number you start with, if you add zero or subtract zero, you'll end up with whatever number you started with. The language "no matter what number" or "whatever number" is a perfect example of an important use of variables: to show something that's always true. For this reason, the **identity property of addition** says that adding zero to a number doesn't change its value, and the **identity property of subtraction** says that subtracting zero from a number doesn't change its value.

When it comes to addition and subtraction, zero has no effect on a number's value. Can you think of a number that has no effect on a number's value when you multiply by it or divide by it? If you are thinking of the number 1, that's correct! When you multiply any number by 1, you do not alter its value; when you divide any number by 1, you do not alter its value. For this reason, the **identity property of multiplication** says that multiplying a number by 1 doesn't change its value, and the **identity property of division** says that dividing a number by 1 doesn't change its value.

Instead of writing out the identity properties in words, you could write them as, "for any number x, it is always true that $x+0=x$, $x-0=x$, $1x=x$, and $\frac{x}{1}=x$." You can plug in any number you want for that variable, and the equation will still be true. Because the identity properties are true properties of numbers, they can be expressed with a variable.

Associative Property

You know from dealing with numbers that where you use parentheses to group numbers in a multiplication problem or in an addition problem has no effect on the truth of the equation. For example, you know that $(3+4+5)=(3+4)+5$ and $(3)(4\times5)=(3\times4)\times5$. And because this is true about all numbers, it's true about variables. Therefore, you know that $(xy)z=x(yz)$ and that $x+(y+z)=(x+y)+z$, no matter what numbers you plug in for the variables.

Why is this important? Well, up until now you've been working on equations that take only one step to isolate the variable. But you're going to start having to solve equations where you'll need to do more than one step to isolate the variable. For example, let's say you're asked to simplify $x+(2+x)=4$, and you want to do so using algebra. Right now, $(2+x)$ is grouped together in parentheses, so it seems like you can't break it up, which brings you to a standstill because you can't simplify $(2+x)$. But, since you know the associative property is true, you know that parentheses don't have any effect here, since $x+(2+x)=x+2+x$. Now you can simplify $x+2+x=4$.

The commutative property tells you that you can rearrange terms that are being added in any order you want without changing the result, so you can rewrite this equation as $x+x+2=4$. Now you can combine like terms, rewriting the equation as $2x+2=4$. You're still not done: to get x by itself, you should subtract 2 from both sides, leaving you with $2x=2$. Now this looks like an equation you know how to solve! To isolate the variable, you divide both sides by 2, and now you know that $x=1$.

You can use these properties to rewrite expressions or equations in ways that are more helpful to you. You don't really have any need to rephrase expressions or equations that are all in terms of numbers, but when variables are involved, rephrasing can really help.

Distributive Property

When you learned about the distributive property with numbers, you discovered that, for example, $4(1+3)$ is equal to $(4)(1)+(4)(3)$. You can do out the math and see that this will be true no matter what numbers you use. Adding together 1 and 3 and then taking 4 groups of their sum is the same as taking 4 groups of 1 and adding them to 4 groups of 3. You can demonstrate this fact through a formula that expresses the distributive property: $a(b+c)=ab+ac$. Notice that you can choose any number you want for *a*, for *b*, and for *c*, and this formula will still be true. It's a property of numbers, so it's true for all numbers.

Now you can see why the distributive property is helpful. With numbers, it doesn't really help you, but with variables, it might. Say you're asked to solve the equation $2(x+4)=3x$. You want to deal with the parentheses first, but you can't because x and 4 are not like terms. What you *can* do is take that 2 and, according to the distributive property, distribute it to each of the terms in the parentheses. In other words, you know because of the distributive property that you can rewrite this equation as $2x+8=3x$.

Now your equation is much easier to work with. In order to solve, you need to combine like terms, which means you should subtract $2x$ from both sides of the equation. Once you do that, you'll have rewritten the equation as $8=x$. The equation is now solved, because you know the value of x.

This is a big function of the distributive property: when you multiply a monomial by a polynomial, you can distribute that monomial and multiply it by each term in the polynomial. Those are important words, but all they do is spell out the distributive property: when you have some numbers that are being added or subtracted, and you want to multiply the sum of their difference by another number, you can instead multiply *each* of them by that other number, and *then* do your addition or subtraction.

When you are multiplying a bunch of terms being added or subtracted (a polynomial) by a single term (a monomial), you can always distribute that monomial to each of the terms in the polynomial. It doesn't matter if the multiplication is written with parentheses or a multiplication sign: $a(b+c) = a \times (b+c) = ab + ac$.

Take a look at one more example. What if you're asked to solve $3(4-x) = x$? You *could* divide both sides by 3 first, but that would give you a bunch of fractions to deal with. Instead, now that you know the distributive property, you know how to multiply a monomial by a polynomial: you can distribute. So you can rewrite the equation as $12-3x = x$. In order to solve, you want to group those x terms together, and it seems like the easiest way to do that would be to add $3x$ to both sides. Once you do that, you have a new equation that tells you $12 = 4x$. You can see that, to isolate the variable, you need to divide both sides of the equation by 4, which leaves you with $3 = x$. Solved!

Adding and Subtracting Polynomials

Another way you can use these properties is to help you figure out what happens when you add or subtract polynomials.

Adding Polynomials

Start by reviewing the rule for adding polynomials, which you might recognize from the associative property: when you add two polynomials, add everything in the two polynomials.

Take a look at the rule in action. Say you're asked to add $(3+x)+(x+2)$. These are two polynomials being added together. What do you do? Well, addition has an associative property, which means the parentheses don't matter at all. So, you can just rewrite the expression as $3+x+x+2$. Now you can follow your rules for simplifying and combine like terms, leaving you with $5+2x$. That's it! You can't simplify any further.

Subtracting Polynomials

What about subtracting polynomials? Say you were asked to subtract $(3+x)-(x+2)$. Well, you know that subtraction does *not* have an associative property, so you can't just ignore the parentheses. That doesn't mean there's nothing you can do. In this expression, the subtraction is telling you to subtract everything in the second set of parentheses. So, you can do exactly that: subtract everything in the second polynomial. You can change this expression to $3+x-x-2$, which can be simplified to 1. If you prefer, you can think of this rule as part of the distributive property. Since subtracting is the same as adding a negative, $(3+x)-(x+2)$ is the same as $(3+x)+(-1)(x+2)$. If you wanted, you could distribute that –1 through the second polynomial and, as you know, it would change the sign of every term in the polynomial. Therefore, when you subtract each term in the second polynomial, you're distributing the negative sign to each term.

When you subtract a polynomial, change the sign of every term in that polynomial. All addition becomes subtraction, and all subtraction becomes addition. Then, once you've distributed the −1 to all terms, you can ignore the parentheses and proceed as normal.

Take a look at just one more example to make sure you've got it. What if you're asked to solve for x in the equation $(2\text{x}+2)-(\text{x}-4)=6$? Okay, well, you're subtracting a polynomial, which means you can distribute that subtraction sign through the second polynomial. Once you do that, the now-positive x will get subtracted. The –4 will be subtracted as well. Keep in mind that when you subtract a negative, it's the same as adding a positive, so you're able to change the signs on all the terms in the second polynomial and rewrite this equation as $2x+2-x+4=6$.

Even though subtraction does not have a commutative property, addition does. As long as you keep the subtraction sign with the term that follows it, you will be okay. Basically, you're treating −x as if it were +(−x) and then moving the negative term wherever you need it to be.

Now you can rearrange your terms. Put the like terms together so you are left with $2x - x + 2 + 4 = 6$. Next, combine like terms and rewrite this equation as $x + 6 = 6$. At this point, you've got a one-step equation, and to solve it, you need to isolate the variable. Subtract 6 from both sides, and you're left with $x = 0$. Solved!

If you want to double-check your answer to an equation, plug your solution in for the variable and make sure it comes out to be true. In the above example, you found that $x = 0$. If you plug in 0 for x in the original equation, you get $6 = 6$. That's true, so your solution must be right!

Order of Operations—in Reverse!

Sometimes, you can do equations that require more than one step without learning anything new. For example, if you're simplifying $\frac{8}{x} = 2$, you can start by doing the only thing there is to do: multiply both sides by x to get rid of the fraction. When you do that, you'll have $8 = 2x$, a one-step equation. You know what to do to isolate the variable—divide both sides by 2, and rewrite the equation as $4 = x$.

Normally, you write an equation that gives the value of a variable with the variable on the left and the number on the right, like $x = 4$. It has the exact same meaning as $4 = x$, but it's just more common to write the variable on the left.

Sometimes it can be confusing to decide which step to do first. For example, if you're told that $\frac{x+2}{4} = 1$, what do you do first? Well, first of all, when you have addition or subtraction in the numerator or denominator as part of a fraction, you should act as if it's surrounded by parentheses, like this: $\frac{(x+2)}{4} = 1$. You have to do this because the entire sum or difference is the value being divided or dividing something, so it acts as if it's in parentheses.

Then, the basic guiding principle is that you need to do PEMDAS in reverse. In other words, you need to undo whatever is being done, starting with the final thing that was done. Think of it this way: if you put an orange in a box, and then you pack an apple on top of it, and then you pack a cookie on top of that, what do you do when you want to isolate the orange? You take out the cookie first; then you take out the apple.

When you're trying to undo something in an equation, you're going to deal with addition and subtraction first, then with multiplication and division, then with exponents and roots, and then with parentheses. Of course if there's something you can simplify on one side of the equation, you can follow PEMDAS as usual.

It can really help to group together your terms with parentheses so that you can see how numbers "travel" together, or are part of one term. Grouping also helps you see when you can simplify on one side of the equation before you start moving numbers to the other side of the equation.

Equations with More than One Step

Everything you've learned so far has prepared you to solve equations that require more than one step. And that's good, because when you can do that, you're pretty much ready for whatever algebra may throw at you. Once you know what you can do and what order you can do it in, it doesn't matter if there are two steps or 100. If you follow the rules, you'll get to the correct answer.

Examples of Equations with More than One Step

Say you're asked to solve for x in the equation $2x \div 4 + 3 - x = 8$. It looks terrifying, but you know you can do it as long as you go one step at a time. Since you're using PEMDAS in reverse, start by undoing any addition or subtraction. You see that you have $+3 - x$ on the left side of the equation, so you're going to move it over to the other side by subtracting the 3 and adding the x. Remember, when you do something to one side of the equation, you have to do the same thing to the other side of the equation. At this point, you are left with $2x \div 4 = 8 - 3 + x$. Now you can combine the $8 - 3$ on the right side of the equation and take one further step toward simplification, leaving you with $2x \div 4 = 5 + x$. You have only one term on the left side, which you can rewrite to rephrase your equation as $\frac{2x}{4} = 5 + x$. If you notice, you can reduce that fraction and change this equation to $\frac{x}{2} = 5 + x$.

Now you've got $\frac{x}{2} = 5 + x$. What do you do next? You know that in order to solve the equation, you have to combine like terms, so you want to get all the *x*s on the same side of the equal sign. You need to move that x on the right over to the left side. Since it is added, you have to undo it by subtracting. Once you subtract an x from both sides, you get $\frac{x}{2} - x = 5$. You're almost there, because now you have two like terms that are easily combined. One of the terms is a fraction, which means that in order to combine them, you need to make both terms into fractions with common denominators.

The identity property tells you that you can rewrite x as $\frac{x}{1}$, and the rules of fractions tell you that you can rewrite that as $\frac{2x}{2}$. So now you have $\frac{x}{2} - \frac{2x}{2} = 5$. You already know that when you subtract fractions with a common denominator, you leave the denominator alone and subtract the numerators—you end up with $\frac{x - 2x}{2} = 5$. This is when you simplify by combining like terms and rewrite the equation as $\frac{-x}{2} = 5$.

Almost done! You can get rid of that 2 in the denominator by multiplying both sides by 2, and you can get rid of that negative sign by multiplying both sides by –1. So you can multiply both sides by –2 to finally find that $x = -10$.

The trick here is to take solving equations one step at a time. Even the best math students have to do these steps one at a time. Be careful, take one permitted step at a time, and you'll get to the right answer. When in doubt, write it out!

Chapter 15 Exercises

Decide whether each of the following illustrates the distributive, the commutative, the associative, or the identity property.

1. $x(x+1) = x^2 + x$
2. $x = 1x$
3. $x + 2 = 2 + x$
4. $x \bullet (y + z) = x \bullet y + x \bullet z$
5. $-1x = -x$
6. $x - 4 = -4 + x$
7. $(x + y) + z = (x + y) + z$
8. $\frac{x}{1} = x$

Solve the following equations for x.

1. $x = 3x$
2. $(3)(x-5) = 60$
3. $x - 4 \div 2 = 3$
4. $4 \div x - 2 = 3$
5. $x \div 2 + 4 = 8$
6. $2x + 4 = 8$
7. $\frac{x+2}{4} = 5$
8. $(x-2) \div 4 = 8$
9. $2(x+3) = x$
10. $-2(x+3) = x$
11. $-x = 4$
12. $2 - x = 3 + x$
13. $(3 + x) \div (3 - x) = 1$
14. $(6 - x) \bullet 2 = x$
15. $4 + x + (3 + x) = 21$
16. $4 + 2x - (3 + x) = 21$
17. $-(2 + x - 4) = x$
18. $4(x-1) = 8$
19. $3 + \frac{2+x}{4} = 7$
20. $1 - \frac{5}{x+2} = 2$

CHAPTER 16

Systems of Equations

Up until now, most of the equations you've been working with have had one variable. That's because to solve for one variable, you usually only need one equation. But what happens when you have two variables? Well, then you need two equations. This chapter will cover how to deal with solving a system of equations, which is really just a group of more than one equation.

More than One Variable, More than One Equation

The next definition you should learn is for linear equations. A **linear equation** is an equation that has two different variables, neither of which is raised to an exponent. So for example, $y = x$ is a linear equation, and so is $y = 2x + 3$.

Once you raise a variable to an exponent, you don't have a linear equation anymore. Why? The word *linear* means "line." Linear equations extend in a line: for every unit you add to x, y gets the same amount bigger (or smaller, depending on the equation) by the same amount that it changes with every other such change in x. Once you add exponents to the mix, this no longer happens. Instead, you start dealing with exponential growth (or exponential decay), which you learned about chapters ago.

With only one linear equation, you can simplify, but you can't solve for anything. Take the simple example of the equation $y = x$. You can do a lot of things to this equation, but none of those things will give you a single value for x or y. All you have is the relationship between x and y. If x is 2, so is y. If x is 5, so is y. You cannot solve for either variable.

To solve for two variables, you're going to need two linear equations.

Substitution Basics

So *how* do you solve for two variables when you have two equations? First, it's important to understand that when you're given two or more equations to use together, a variable must stand for the same value in all equations. A group of two or more equations is called a **system of equations**.

Within a system of equations, all variables must keep their values. So if you have a system of equations using *x* and *y*, you are using the same *x* in both equations, and the same *y* in both equations.

There are two ways to solve a system of equations, but the one that is most commonly used is substitution. The principle behind substitution is that you can make an equation with only one variable if you use the relationship in the other equation to replace one variable with the other. Sound complicated? It is best explained with an example. Say that you are asked to "solve for x and y" and are given the two equations $x = y$ and $2x = y + 10$. So you have two linear equations, which means you probably have enough information to solve for both variables.

ALERT

It is possible for two equations to actually be the same equation, if they're not fully simplified. For example, if you were given the two equations $x = y$ and $2x = 2y$, you still can't solve for the variables because the second equation can be reduced to $x = y$. These are really just two ways of writing the same equation.

The first equation tells you $x = y$. That means those two things are equal. Because they have the same value, you can use them interchangeably. And because of that, you can replace x with y in the second equation. (You could also replace y with x and end up with the exact same results.) The second equation, when you make that substitution, changes from $2x = y + 10$ to $2y = y + 10$. Now you have an equation with only one variable, so you can follow the steps to solve for the variable.

Start by combining like terms, which you do by subtracting y from both sides, and you're left with $y = 10$. Now that you've solved for y, you can plug 10 in for y, because you've proven that these two things have the same value. It doesn't matter which equation you plug 10 into. If you plug it into the first, you now know that $x = 10$. If you plug it into the second, you know that $2x = 20$, which tells you that $x = 10$. Since you can plug the solution into either equation, you should plug it into whichever is simpler; in this case, that's the first equation. Either way, you now know that $x = 10$ and $y = 10$, and you're finished solving.

Substitution in Practice

Substitution is an important skill to build. Sometimes you'll be asked to solve for both variables. Sometimes you'll only be asked to solve for one. You follow the same steps in both cases, but you can take a little shortcut when you're asked to solve for only one variable.

First, look at an example where you are asked to solve for both variables: Solve for x and y given the equations $2y = 4x + 2$ and $x + y = 10$. What should you do?

1. **Select the easier question.** Since you're going to solve for both variables, it doesn't matter which one you start with. $x + y = 10$ looks easier, because the variables don't have any coefficients.

A **coefficient** is a number that is multiplied by a variable. So, for example, in the term 3z, the coefficient is 3. If a variable has no coefficient, it's the same as having a coefficient of 1.

2. **Isolate the variable in that equation.** In this case, you're going to be solving for both variables, so you can decide to isolate either one. Do whichever one is easier. In this case, you can simplify $x + y = 10$ by subtracting y from both sides, which isolates x and leaves you with $x = 10 - y$.
3. **Plug the new expression into the other equation to solve for one variable.** Now that you know $x = 10 - y$, you can take $10 - y$ and plug it in for x in the other equation. Go to the second equation of $2y = 4x + 2$ and plug in $10 - y$ every time you see an x. Once you do that, you'll have $2y = 4(10 - y) + 2$.

When you plug in an expression for a variable, be sure to plug that expression in with parentheses. The whole expression is equal to the variable, so you need to replace the variable with *everything* in the expression. You do that by using parentheses.

4. **Solve the new equation for its one variable.** Now that you have a one-variable equation, you can solve for *y*. Follow the rules of solving equations and distribute the 4 to the terms in the parentheses; then combine like terms on the right side of the equation to get $2y = 42 - 4y$. Combine like terms to get $6y = 42$, and then isolate the variable by dividing both sides by 6. You now know that $y = 7$.
5. **Plug your solution into either original equation to solve for the other variable.** The easier equation is probably $x + y = 10$, so plug 7 in for *y*, and now you know $x + 7 = 10$. You can solve this equation to determine that $x = 3$. You've solved this system of equations! You found that $x = 3$ and $y = 7$.
6. **If you want, double-check by plugging your solutions into the original equations and making sure they are both true.** $2(7) = 4(3) + 2$ and $3 + 7 = 10$ are both true, so you know you've got the right answers.

What if you're only asked to solve for one variable? Sometimes, the problem will make your life easy by telling you the value of the other variable. For example, what if a problem says, "In the equation $y = 4x - 1$, what does *y* equal when *x* equals 3?" That's the same as being asked to solve for *y* given the two equations $y = 4x - 1$ and $x = 3$. You can take that 3 and plug it into the first equation. This gets you $y = 4(3) - 1$, which you can simplify to $y = 11$.

Solving for One Variable

Sometimes you'll be given two equations and asked to solve for only one variable. For example, you might be asked, "Solve for *x* when $x + y = 3$ and $x - y = 7$." Well, you could follow the process above, and you'd get the right answer, but you want to make sure you don't waste any time solving for *y*, because you don't need to solve for *y* in this problem. Thus, if you want, you can amend the steps a little to take the shortest route possible to the answer. Here are the steps in solving a system of equations for only one variable:

1. **Isolate the variable you *don't* want to solve for.** In this example, you're being asked to solve for *x*. Because of that, you want to rearrange one of the equations so that *y* is isolated. The first equation looks easier, so you're going to rearrange it to be $y = 3 - x$.

2. **Plug the new expression for the variable you don't want into the other equations.** Now that you know $y = 3 - x$, you can plug $(3 - x)$ in for for y in the second equation. Once you do that, you'll have $x - (3 - x) = 7$.
3. **Solve the equation for the variable you want.** You have one equation with one variable, so you can solve for that variable. To solve $x - (3 - x) = 7$, you would first distribute the negative among the terms in the parentheses and rewrite the equation as $x - 3 + x = 7$. When you combine like terms, you get $2x - 3 = 7$. Now, add 3 to both sides to isolate the variable, which gives you $2x = 10$. Solve by dividing both sides by 2, and you're left with $x = 5$.

Combining Systems of Equations

You now know how to solve a system of equations by substitution. That's probably the method you'll use most, if not all of the time. Why? It always works, and most of the time it's also the shortest route to solving for both variables. But you need to learn how to combine a system of equations in case you're asked to do it . . . and also because sometimes it's a great shortcut.

Say you're asked to solve for x and y given the equations $x + y = 3$ and $x - y = 11$. You could go through the whole process of substitution, and it would work. But there's something else you can do: you can add the equations together.

You can always add two equations together and get another true equation. For example, if you add $3 = 3$ to $7 = 7$, you'll end up with $10 = 10$, which is true. That's because if two values are equal and you add the same amount to both of those values, you'll still have two values that are equal.

If $x + y = 3$, then $x + y$ has the same value as 3. They are the same. They would have the same "weight" on a scale. And you know that you can add any amount you want to both sides of an equation, as long as you add the same amount to both sides. Since this equation tells you that $x+y$ is the same amount as 3, you can add one of these expressions to one side of an equation, and the other expression to the other side—and that new equation will still be true.

Because of this, you can, if you think it will be easier, solve a system of equations in the following way:

1. **Stack the equations one on top of the other.** Align the equal signs. So you might put $x+y=3$ on top of $x-y=11$, with the equal signs lined up.
2. **Add the left sides of the two equations together to get their sum.** In this example, you would add $x+y$ and $x-y$, giving you $x+y+x-y$. Notice what happens? When you try to simplify, you put the like terms together, giving you $x+x+y-y$. The x terms combine to give you $2x$. The y terms actually disappear: $y-y=0$. So your left side of the equation is just $2x$.
3. **Add the right sides of the two equations together to get their sum.** In this example, you would add $3+11$, and the right side of your equation becomes 14.
4. **Solve your new equation.** The new equation is $2x=14$, which you can solve to figure out that $x=7$.

If you wanted to, you could take the solution for x and plug it back in to either of the original equations to solve for y.

You can see that you would probably only choose this method if it caused one of the variables to cancel out completely; otherwise, it will just give you a third equation with two variables, and won't be much of a help. You can use this method when one of the variables has the same coefficient in both equations. If that coefficient is positive in one equation and negative in the other (as it was here), adding the equations will cause the variables to cancel each other out. If the coefficient is positive in both or negative in both, you'll have to subtract the equations instead of adding them. It doesn't matter which one you subtract from the other—you should follow the same steps as given above, but with subtraction instead of addition.

Introduction to Formulas

Sometimes you are given an equation with more than one variable, but neither of the variables stands for one particular number. The equation is not intended to be solved by finding the only value that x and y could possibly be. Instead, the equation is intended to show a relationship; it's intended to answer the question, "When x has a certain value, what will be the value of y?" An equation you use in this way is called a formula.

> You've seen a lot of formulas in your life. Anything that shows the relationship of two variables is a formula. So, "for every cup of flour, add two cups of sugar" is a formula because it gives you the ratio relationship of two variables: the amount of flour and the amount of sugar.

It's best to look at an example of a formula to see what it means, how it works, and how you might be asked to use it. Imagine that you're told $y=2x+3$, where x is the number of hours spent painting and y is the square footage painted. What does this mean? It means that you have a formula that shows the relationship between time spent painting and square footage painted. If you know how many hours have been spent painting, the formula will tell you how much square footage has been painted. Also, if you know how much square footage has been painted, you will know how much time has been spent painting.

If someone asks you how many square feet Jane will be able to paint if she spends three hours painting, now that you know the formula, you are able to answer the question. You've been told that 3 is the number of hours spent painting. You know that x in your formula stands for the number of hours spent painting. So, you know that, in this particular instance, $x=3$. Armed with that information, you can plug in 3 for x in the formula and learn that when $x=3$, $y=9$. Therefore, if Jane spends three hours painting, she will be able to paint nine square feet.

You will use formulas a lot in math to show relationships that stay the same, regardless of what the numbers are. For example, the area of a rectangle can always be represented by the equation $a=lw$, where a represents the area of the rectangle, l represents the length, and w represents the width. This relationship is always true in a rectangle, so if you are given any two of these three numbers, you can find the third.

The number properties you just learned also give you formulas. It's true for any real numbers that $a(b+c)=ab+ac$. This statement is called the distributive property, but it's really just a formula. It's a formula that stays true for any real numbers: no matter what you plug in for a, b, and c, this relationship holds true.

Introduction to Functions

Functions are a special kind of formula. For almost all of algebra I and II, you'll be dealing with functions, so it's important to take some time to understand what a function is. Here's a simple definition: a function is an equation that spits out a unique answer for each variable you plug into the equation. For example, look at the equation $y=x+2$. You can plug in any number you want in the place of x, and for any x you plug in, this equation will spit out one unique value for y.

You can think of a function like a machine that does something to a number. For any number you put into the machine, the machine always does the same thing to that number, and then spits out the result.

You can express functions in lots of different ways: pictures, graphs, charts, and formulas. Most functions you'll deal with in math can be represented as a formula, or an equation. Consider an example: $y=x+2$ is a function. When you plug in a value for x, this function tells you what happens to x to get to y, and eventually tells you what y will be. If you put in a 2 for x, the function tells you that y must be 4. In other words, when $x=2$, $y=4$. When you plug in 3 for x, the function tells you that $y=5$. The function doesn't tell you the value of x or y; their values could be anything. But it *does* tell you the *relationship* between x and y.

A function does not tell you what either variable equals. It's an equation with more than one variable that can't be solved in the traditional sense. Instead, it tells you what one variable is in relation to the other one. It lets you solve for one of the variables only when you're given the other one.

You can think of a function as a way of relating two pieces of information. For example, think about the sentence "Bill is two inches taller than Dan." You don't know how tall Bill is, or how tall Dan is, but if you are told how tall Dan is, you can tell how tall Bill is, because you know the relationship of those two numbers.

Functions are equations, and they follow all the rules of equations. You plug in the value you're given, and you can solve for the remaining value. What makes functions confusing for most people is the way they are written. But you're going to see that it's just a new code, a new way of writing something. It doesn't ask you to do anything that you don't already know how to do.

Examples of Functions

Reconsider the function $y = x + 2$. If you wanted to write this same relationship in traditional function notation, you'd write $f(x) = x + 2$. This notation uses a symbol you already know in a new way. You know that parentheses usually show multiplication. For example, in the expression $2(x + y)$, $x + y$ is being multiplied by 2. But functions are set up so that one side of the equation looks like $f(x)$, which you read as "*f* of *x*," which stands for "function of *x*." In this case, *f* isn't a variable, and it is not being multiplied by *x*. You use the notation $f(x)$ to stand for "function of *x*." You can recognize this notation because it stands alone on the left side of an equation.

It is possible to use different letters to stand for "function." When you have more than one function, you often call the first one *f* and the next one *g*. It's also possible to use any variable in the equation, not just *x*. If your equation has *y* as the variable, then it would be a function of *y*, or *f*(*y*).

So in the function $f(x)=x+2$, what do you know? You know that the function of *x* is $x+2$. In other words, you know that when you put *x* in the "function machine," its value will increase by two. Whatever value you put in for *x*, the function of *x* will be that value plus 2. Basically, $f(x)=x+2$. Your teacher could give you the function $f(x)=x+2$ and ask, "What is $f(x)$ when $x=4$?" Your teacher is asking, "What will this machine spit out as the answer when you plug in a 4 for *x*?" You can plug in 4 for *x* and see that the machine spits out a 6. So you know that when $x=4$, $f(x)=6$.

This is the exact same thing you did when you had an equation with two variables and were told the value of one of the variables. You plug in for the variable you know and check to see what the equation "spits out" as the right answer for the other variable. You're doing the exact same thing here, only the "other variable" is $f(x)$. So when you are asked, "Given the equation $y=2x-1$, what does *y* equal when $x=5$?", you would solve it by the exact same method as if you were asked, "Given the function $f(x)=2x-1$, what does $f(x)$ equal when $x=5$?" It does not matter how the equation is set up. You plug in the value you've been given for one of the two missing pieces of information, and you're able to solve for the other one.

Chapter 16 Exercises

Use substitution to solve for *y* in the following equations.

1. $xy=8$, $x=2$
2. $x=3$, $\frac{y+x}{2}=8$
3. $\frac{x}{y}=5$, $x=10$
4. $x=1$, $y-1=x$
5. $x+4+\frac{y}{2}=16$, $x=4$

Use substitution to solve for *x* and *y* in the following systems of equations.

1. $x+y=5$, $2x=10+y$
2. $x-y=10$, $x+y=20$

3. $\frac{x}{2} = y$, $x = 1 + y$
4. $3x - y = 8$, $\frac{x+y}{3} = 4$
5. $\frac{16}{x+y} = 8$, $x - y = 0$
6. $2 - y = 10 - x$, $-x = y$
7. $x = 3y$, $3 + 6y = 3x$
8. $y = 3 + x$, $\frac{y}{4} - 2 = -x$
9. $y = 3x$, $\frac{6-x}{2} \bullet 3 = y$

Answer each of the following questions.

1. Given $f(x) = 3x + 9$, what is $f(x)$ when $x = 3$?
2. Given $f(x) = \frac{2}{x} + 4 \bullet 2$, what is $f(x)$ when $x = 2$?
3. Given $f(x) = \frac{1-x}{4}$, what is $f(x)$ when $x = -3$?
4. Given $f(x) = 10 - 2x$, what is $f(x)$ when $x = 1$?
5. Given $f(x) = 4x + \frac{x}{3}$, what is $f(x)$ when $x = 3$?
6. Given $f(x) = \frac{x+3}{x-3}$, what is $f(x)$ when $x = 0$?

Use the formulas given to answer the questions below.

A machine makes bolts and nails. The number of nails it can produce is determined by the formula $n = 400h$, where n stands for the number of nails and h stands for the number of hours. The number of bolts it can produce is determined by the formula $b = 50d$, where b stands for the number of bolts and d stands for dollars spent.

1. How many bolts can be produced for $10?
2. How many nails can be produced in 3 hours?

3. How long will it take, in hours, to produce 4,000 nails?
4. How much will it cost, in dollars, to produce 5,000 bolts?

Kelly saves her money according to two different formulas. For her babysitting money, her savings is determined by the formula $S = \frac{b}{2} + 1$, where S is the amount saved in dollars and b is how much money (in dollars) she earns babysitting. For her tutoring money, her savings is determined by the formula $S = t - 10$, where S is the amount saved in dollars and t is how much money (in dollars) she earns tutoring.

1. How much money will Kelly save from $30 she earns tutoring?
2. How much money will Kelly save from $30 she earns babysitting?
3. How much babysitting money does Kelly have to earn to save $11?
4. How much tutoring money does Kelly have to earn to save $11?

CHAPTER 17

Inequalities

Inequalities are hard to think about, because they aren't complete information. When you are told $x = 5$, you know everything you can possibly know about x. It's 5. Mystery solved. But when you are told $x > 5$, which in words translates to "x is greater than 5," you know *some* things about x, but not everything. This chapter will help you learn what inequalities are, how they work, and how you can "do math" with them. In a lot of ways, they work just like equations, so don't be afraid! You already know how to do most of this stuff already.

Symbols

There are four inequality symbols you need to know. The word *inequality*, as you can tell by looking at it, means "not equal." Inequalities are symbols that show the relationship of two quantities that aren't equal. There are symbols in math to show, when comparing two quantities, that one is greater than the other. The "greater than" symbol is $>$ and the "less than" symbol looks like $<$.

If you would rather think of "bigger" and "smaller" instead of "greater" and "less," you can. Although "bigger" and "smaller" technically refer to size, whereas "greater" and "less" refer to amount, this is a math book, not a grammar book. Whatever way is easiest for you to think about, go for it!

Teachers often explain that you can think of the symbol as an alligator's mouth: the mouth opens toward the bigger number, because the alligator wants to eat the bigger number. The symbol is bigger on one side (the open side) and smaller on the other side (the closed side), so it only makes sense to put the bigger side of the symbol facing the bigger number. So, for example, $5 > 3$ and $3 < 5$. Notice that these two inequalities express exactly the same information, just in different ways. You read the first statement by saying, "five is greater than three," and you read the second statement by saying, "three is less than five."

The other two symbols are not used much with numbers, but they are often used with variables. These are the symbols that stand for "greater than or equal to" and "less than or equal to." The "greater than or equal to" symbol combines the $>$ with $=$ to give you $\geq$. So if you write $x \geq 5$, you're saying that "x is greater than or equal to 5." In other words, x could be 5, or it could be anything bigger than 5, but it can't be smaller than 5. You often see this idea expressed in word problems with the phrase "at least." If you have "at least 10 apples," you can write $a \geq 10$ in order to say that a, the number of apples, is greater than or equal to 10.

Just like the "greater than or equal to" symbol, you have a "less than or equal to" symbol. It combines the $<$ with the $=$ to give you $\leq$. So if you write $x \leq 5$, it means "x is less than or equal to 5." In other words, x could be 5, or it could be anything smaller than 5, but it can't be bigger than 5. You will see this in word problems with phrases such as "no more than" or "up to." If you have "no more than 10 apples," you can write $a \leq 10$ in order to say that *a*, the number of apples, is less than or equal to 10.

Back to the Number Line

Inequalities are okay when all the numbers are positive integers. This is most likely how you first learned inequalities, and they probably made pretty good sense to you. Of course 10 is bigger than 5! You don't even have to think about it—you know it's bigger.

But, as usual with math, things get more complicated when you realize that there are a lot of numbers that aren't positive integers. There are negative numbers. There are fractions. There is zero. It's a lot to think about, and it can make things pretty tricky when it comes to inequalities.

The best way to begin to understand how inequalities relate to all these non-positive-integer numbers is to break out the old number line. Take a look:

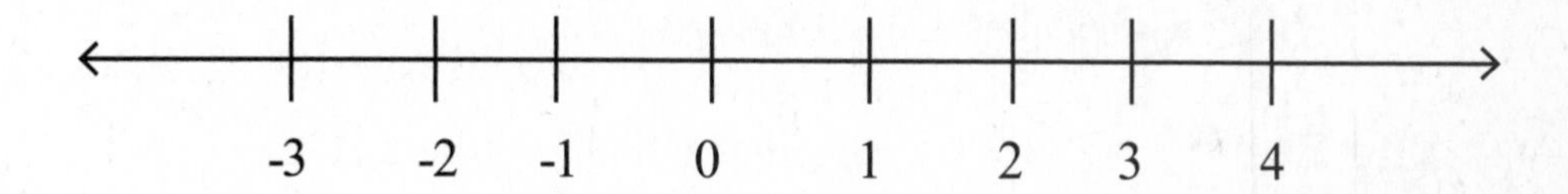

The number line arranges numbers in order of size. It takes whatever numbers you give it and sorts them out so that the biggest number is farthest to the right and the smallest number is farthest to the left. It goes on forever in both directions and contains an infinite amount of numbers. And all of them are in order.

Take a look at what happens on the positive side: the bigger magnitude a number has (the farther it is from zero), the bigger that number is. But now take a look at the negative side: the opposite is true! The bigger magnitude a number has (the farther it is from zero), the *smaller* that number is. So while 2 is greater than 1, –2 is actually smaller than –1.

There are a few ways to remember this concept. If you are a visual learner, seeing it on the number line may help you. If you prefer to memorize facts, you could memorize the fact that negative numbers get smaller the bigger their magnitude is. If you prefer to think in real-world concepts, it helps to think about money. If you assume you'd rather have more money than less money, would you rather have $2 or $1? $2, of course. With money, negative numbers are basically debt, or money you owe. So would you rather have –$1 or –$2? In other words, would you rather owe $1 or $2? You'd rather owe $1; it's a smaller debt, which is actually more money for you. If you have to be in debt, you want it to be as close to zero as possible.

Notice one other thing about the relationship of negative and positive numbers on the number line: any negative number is smaller than any positive number, because all negative numbers are smaller than all positive numbers. So if you're comparing a positive and a negative, the positive will always be bigger, no matter what the values themselves are.

Inequalities Without Variables

The rules of inequalities are best learned by examining inequalities that do not have variables. You already know that you make the inequality sign "open" to the bigger number, but there are a couple of other rules you should know about inequalities before you get into adding, subtracting, multiplying, and dividing.

The Rules

First, you should know that an inequality works sort of like a scale. The inequality compares two quantities, and it tells you which one is bigger, or which one would be heavier on a scale. For example, $100 > 50$ shows you that the left side of the scale is heavier than the right. But, if you wanted, you could spin that scale around on the table, and then the right side would be heavier than the left. It's the same piece of information; you're just looking at it a different way. You can do the same thing when you write an inequality. You can just as easily write $50 < 100$. Just swap the right and left sides of the scale, and turn the inequality symbol around.

You can still do this even when both sides of the inequality are written as expressions. Take a look at the following example: $5+2-3>1-1-5$.

Say that you want to rewrite this inequality so that it has a "less than" sign instead of a "greater than" sign. You want to think of each side as one piece. It might be useful to add parentheses just for you to think about it as

$(5+2-3)>(1-1-5)$. Now you can just swap those two pieces and flip the inequality around. It's just like turning around the scale and looking at it from the other side. In the end, you have $(1-1-5)<(5+2-3)$.

Of course, you can make your life a whole lot easier by simplifying *before* you try to do anything else. Going back to $5+2-3>1-1-5$, you could simplify the left side and the right side and rewrite the inequality as $4>-5$. Then, if you wanted to flip it, your work would be easy! You'd get $-5<4$.

Why Flip?

Why would you bother to flip an inequality? Well, there are a couple of reasons. When you start graphing in algebra class, you might want your inequality to take a particular form. Most importantly, you sometimes have more than one inequality and want to combine them. To do that, you have to have the inequality symbols facing the same way, which sometimes means flipping them around.

One last thing to know about inequalities before you start manipulating them is that they sometimes have more than two parts. For example, you can say $2<3<4<5$. That's true! 2 is less than 3, and 3 is less than 4, and 4 is less than 5. Say you started with these three true statements:

$$2<3$$
$$3<4$$
$$4<5$$

You can combine all this information into one easy-to-read statement by using an inequality with more than two parts. Sometimes you do this in order to set the boundaries of what a number could possibly be. For example, take the inequality $5<x<10$. You know that x is somewhere between 5 and 10. This is a math concept you use a lot in the real world, because a lot of things have a maximum and a minimum. For example, a certain business plan might require at least $1,000 but not be able to use more than $5,000. Inequalities that relate more than two pieces of information are commonly seen in systems of boundaries such as this.

Manipulating Inequalities with Addition and Subtraction

When it comes to addition and subtraction, you are in luck, because inequalities work just like equations. You can add whatever you want, as long as you add the same thing to every quantity in the inequality. That's because doing so is like adding the same amount of weight to both sides of a scale. The total values will go up, but whichever amount was bigger before will still be bigger. As long as you increase all the values being compared by the same amount, their relationship won't change.

Take a look at this principle with numbers. It's true that $7 > 5$. Whatever you add to both sides, their relative magnitude (and the inequality symbol that expresses that relationship) will remain the same. For example:

$$7+2>5+2 \quad \text{(True! } 9>7\text{)}$$

$$7+100>5+100 \quad \text{(True! } 107>105\text{)}$$

If your inequality has more than two parts, remember that the relationship only holds up if you add the same amount to every part, not just the outside ones. For example, $5<6<7$. Then it must be true that $(5+2)<(6+2)<(7+2)$ . . . and it is! $7<8<9$.

Subtraction works in exactly the same way. Take the same amount away from both sides, and whatever was bigger before will still be bigger after. As long as you subtract the same amount from each value in the inequality, your inequality will still hold true. For example, say you're told $10>6$.

When you take 2 away from both values, they both get 2 smaller, but their relationship stays the same: whichever one was bigger before is still bigger, even though they've both changed in size. So, because $10>6$, it must be true that $(10-2)>(6-2)$, and it is! You know that $8>4$.

There's one other very cool thing you can do with inequalities. When you have an equation, you have two quantities that are equal to one another. So, you can take two equations and add them together. For example, consider these two separate equations:

$$5=3+2$$
$$6=7-1$$

In each equation, the quantities on each side of the equal sign are equal to one another. Because of that, you can add the two equations together, and you will get another true equation. Add up the left sides of the equations: $5+6=11$. Add up the right sides of the equations: $3+2+7-1=11$. By adding the equations together, you've created a third true equation: $11=11$.

FACT

If you take the same number and add it to both sides of an equation, you'll always get a new true equation, because you've increased both sides of the equation by the same amount. It's like taking two balanced scales, combining the quantities from the two left sides on the left side of a new, third scale, and then combining the quantities from the two right sides on the right side of the third scale. When all that has been done, the third scale will also balance.

And inequalities can be added together in the same way, without changing the relationship between the left and right sides or changing the inequality symbol used to express that relationship. Take a look at the following two inequalities as an example of what happens when you add two inequalities together. (You could use any two inequalities as an example, because this always works.)

$$6>5$$

$$3>0$$

In the first inequality, 6 is bigger than 5. In the second, 3 is bigger than 0. Picture two scales, each of which is tipped toward the left number. If you were to add up the two heavy sides of the scale, their combined weight would definitely be bigger than the two light sides of the scale put together. Using this example, you can see that it's true: $6+3$ (the heavy sides) is greater than $5+0$ (the lighter sides) because 9 is bigger than 5.

While you *can* subtract one equation from another, you *cannot* subtract one inequality from another. Sometimes it comes out true, and sometimes it comes out false. This is one of the only differences between inequalities and equations, so be careful! You can subtract anything you want from both sides of an inequality, but you *cannot* subtract one inequality from another.

Manipulating Inequalities with Multiplication and Division

Manipulating inequalities with multiplication or division is, for the most part, exactly like manipulating equations with multiplication or division. Just like with an equation, you can multiply both sides of an inequality by the same number, and the inequality will still be true. That should make sense if you go back to the scale analogy. For example, take the inequality $5 > 3$.

This inequality shows you that 5 is bigger than 3. In other words, 5 is heavier on the scale than 3 is. If you multiply both sides by the same thing, the values will change, but the relationship won't. Why? Let's say that you multiply both sides of the inequality by 4 to get $(5 \times 4) > (3 \times 4)$.

You now have four of the heavier thing on the left, and four of the lighter thing on the right. Of course four of the heavier thing will still be heavier than four of the lighter thing! So as long as you multiply both sides by the same value, the new inequality will still be true.

The same goes for division. Let's say that you're given $20 > 12$. You can divide both sides by anything you want, and this inequality will still be true. Why? Because if 20 is heavier than 12, and you divide each of these values into the same number of groups, each group on the heavier side will be heavier than each group on the lighter side.

As an example, divide both sides by 2. Now you have $10 > 6$. That's true! Why? Well, when you divide something by 2, you're really just taking half of it. And half of a heavier object is still going to be heavier than half of a lighter object. The key here is to do the same thing to both sides. As long as you multiply or divide both sides of the inequality by the same value, the relationship will stay the same.

Remember how, at the beginning of this section, you were told that manipulating inequalities by multiplying or dividing was *for the most part* the same as manipulating equations by multiplying or dividing? That's because there's one big difference you have to keep in mind. It has to do with negative numbers.

The Rule

The rule is that when you multiply or divide both sides of an inequality by a negative number, you have to flip the inequality. So, if you were to multiply both sides of an inequality by a negative number, you would change a "greater than" sign into a "less than" sign, and you would change a "less than" sign into a "greater than" sign.

Why? Remember your number line. With positive numbers, numbers with a bigger magnitude also have a bigger size. But with negative numbers, numbers with a bigger magnitude have a *smaller* size. Inequalities show a relationship of size, so when numbers change their sign, their size relationship changes.

While you can do *almost* anything to an inequality and keep the relationship the same, if you multiply or divide both sides of a inequality by a negative number, you have to remember to flip the sign. The relationship will still be known—it will just be the opposite of what it was before.

Take a look. Let's start with a simple inequality, such as $5 > 3$. What if you multiply both sides by –1? The left side will turn into –5, and the right side will turn into –3. Which one of these is bigger? –3 is actually bigger than –5, because it's farther to the right on the number line. Negative numbers get bigger the smaller their magnitude (or distance from zero) is. So when you multiply by a negative number, you have to change the direction of the inequality symbol. Therefore, if you multiply both sides of the inequality by –1, you end up with $-5 < -3$. You have to turn the inequality symbol around to make this inequality true.

Notice that the same rule applies if you divide by –1. To keep the inequality true after dividing both sides by –1, you have to change the direction of the inequality symbol as your final step.

Inequalities with Variables

Now that you've learned all the rules for manipulating inequalities by looking at examples with numbers, you can apply all these same rules when variables are involved. If a rule works with numbers, it works with variables. The trick to algebra is to always understand the rules with numbers, follow those same rules with variables, and then practice with variables enough that you get comfortable with them. It's not glamorous, magical, or immediate, but it does work, and it does make algebra stick.

Examples of Inequalities with Variables

Start with an inequality with one variable, such as $x > 5$. In this inequality, you don't know what x is, but you know it has a value bigger than 5. So what else do you know? In other words, how can you manipulate this inequality so that it tells you something else that's true?

Well, you can add or subtract whatever you want from both sides, and it will still be true. If you picture the scale again, when you add the same amount to both sides, whichever side was heavier before will still be heavier after. When you subtract the same amount from both sides, whichever side was heavier before will still be heavier after. So here are some things that must be true:

$$x+2>7$$

$$x-2>3$$

$$x+15>20$$

$$x-10>-5$$

Notice that, if you want, you can simplify all of the examples above, and they'd take you right back to the original inequality. Check it out:

$x+2>7$ → Subtract 2 from both sides to get x alone → $x>5$

$x-2>3$ → Add 2 to both sides to get x alone → $x>5$

$x+15>20$ → Subtract 15 from both sides to get x alone → $x>5$
$x-10>-5$ → Add 10 to both sides to get x alone → $x>5$

You can also multiply or divide both sides by whatever value you want, but if that value is negative, you have to change the direction of the inequality. So if you start again with $x>5$, here are some things you can do:

$x>5$ → Multiply both sides by 2 → $2x>10$
$2x>10$ → Divide both sides by 2 → $x>5$
$x>5$ → Multiply both sides by –2 → Don't forget to flip the inequality! → $-2x<-10$
$-2x<-10$ → Divide both sides by –2 → Don't forget to flip the inequality! → $x>5$

Everything you can do with inequalities that contain only numbers, you can do with inequalities that have variables in them.

For example, consider this inequality that relates two variables: $x>y$. You don't know what x is, and you don't know what y is, but you know x is bigger than y. So you can add anything you want to both sides, subtract anything you want from both sides, multiply both sides by anything you want, or divide both sides by anything you want, and the inequality will still give you a true statement. The only special hitch you have to remember is that if you multiply or divide both sides by a negative number, you have to flip the direction of the inequality. So given that $x>y$, here are some other things that must be true:

$$x+2>y+2$$
$$x-5>y-5$$
$$\frac{x}{2}>\frac{y}{2}$$
$$-2x<-2y$$
$$-x<-y$$

Unlike equations, you can't use an inequality to solve for a variable. That's because inequalities don't tell you what a variable equals; they just tell you what it's bigger than or smaller than. Often you'll be told to *simplify* the inequality, which means to isolate the variable on one side of the inequality (whenever possible), just as you'd do if you were solving an equation.

Even though the directions for inequality problems usually say "simplify" instead of "solve," you should follow the same process you use when solving an equation: isolate the variable by undoing whatever's being done to it.

Chapter 17 Exercises

Fill in each blank with either > or <.

1. 3__5
2. −3__−5
3. −3__2
4. −10__−100
5. 1.5__1.4
6. −1.5__−1.4

Simplify the inequality.

1. $x+2<15$
2. $y-3<8$
3. $x+14>-3$
4. $y-12<-14$
5. $2x<4$
6. $x\div 3>21$
7. $4y>10$
8. $\frac{x}{2}<14$
9. $-3y<1$
10. $x>-1$
11. $-b>-5$
12. $\frac{b}{-1}>5$
13. $\frac{-b}{2}<-5$

CHAPTER 18

Absolute Value

The nice thing about absolute value is that it's pretty self-contained. It's a concept you need to understand and a skill that will keep resurfacing in math. This chapter will cover what absolute value really means and how to solve absolute value equations.

Back to the Number Line . . . Again!

You've actually already learned a little bit about absolute value back near the beginning of this book, when you learned about magnitude. The magnitude of a number is the same as its absolute value: they both tell you the distance of that number from zero.

Check out the number line below:

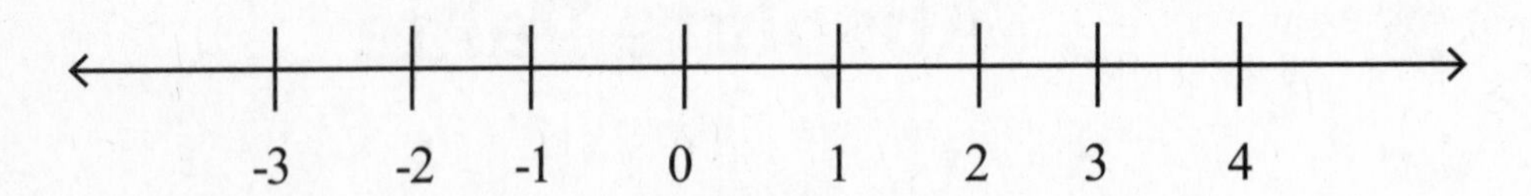

Notice that for the positive numbers, their distance from zero is simply their value. So the absolute value of 3 is 3, for example. For negative numbers, the absolute value is similar to the value of the number, except it is positive. For example, the absolute value of –3 is 3.

Why is the absolute value always positive?
The absolute value tells you the distance of a number from zero, and distances are always positive. If you walk one mile to the store, your walk back in the other direction is still one mile. There's no such thing as "negative one mile." Distances exist in the real world, so they can't be negative numbers.

Rules of Absolute Values

The symbol for absolute value is two vertical bars around a number. Now that you're good at using a variable to represent any number, you can try using a variable to show the formula. For any positive number x, $|x| = x$. You would read that equation as, "The absolute value of x is equal to x." In other words, the absolute value of any positive number is equal to itself. The same formula holds true when $x = 0$.

When a formula or rule starts out by saying "for any positive number x," that means, "here comes a statement that will be true for any positive number that you plug in for x."

For negative numbers, as you know from the number line, things are going to be a little different. In conversational terms, when you take the absolute value of a negative number, you simply change the sign to positive. In math terms, you might say: for any negative number *y*, $|y| = -y$.

Whoa, whoa, wait a second. You just learned that the absolute value was always going to end up being positive, so how does $|y| = -y$ work? Well, remember that $-y$ doesn't have to be a negative number. Putting a negative sign in front of a variable is the same as multiplying it by –1. In other words, it just changes the sign of the variable. So if *y* were positive, $-y$ would be negative. If *y* were negative, though, $-y$ would be positive. In this case, you know from the constraint given at the beginning of the formula that *y* is negative: this formula holds true only when *y* is a negative number. In that case, $-y$ is going to be positive.

Be careful of negative signs in front of variables! They don't necessarily mean negative numbers, but they do change the sign of the variable. So anytime the variable stands for a negative number, putting a negative sign in front of that variable will make it positive.

Negatives and Positives with Variables

Absolute value is an expression of the size of a number in positive terms. When you're dealing with numbers, there's no question whether a number is positive or negative, but when you start dealing with variables, things get a little tricky.

How will you know whether a variable is positive or negative? Here's how:

1. **The problem explicitly tells you.** Sometimes a problem or problem set will start by pointing out that "x is positive" or "x is negative." When that happens, pay attention to the directions you've been given!
2. **The problem is a word problem with real-world constraints.** *Real-world constraints* is a term used to describe things that must be true just because the problem exists in the real world. Remember *real* numbers? Those numbers exist in the real world. So if in your word problem, x represents the number of dogs, it can't be negative, because you can't have negative dogs.
3. **The problem creates a situation that's only possible with positive numbers.** If you're given, for instance, the fact that $x^2 = y$, you know that y is not negative. No matter whether you plug in a negative or positive number for x, you can never have y come out to be negative. In another example, if you're given $\sqrt{x} = y$, you know that x cannot be negative, because it's impossible to take the square root of a negative number.

Absolute Value with Rational Numbers

What if you're asked to simplify $|-4|+3$? What do you? Well, you want to simplify the absolute value first. You know that the absolute value of a number is just the positive version of that number, so the absolute value of –4 is 4. Thus, you can replace $|-4|$ with 4 and rewrite the expression as $4+3$, which can be further simplified to 7.

What if the expression is a little trickier, and you are asked to simplify $|4+2| \div 3$? Although the absolute value bars have a different meaning than parentheses, for PEMDAS purposes, you should treat them like parentheses. The absolute value bars group the terms inside them together in way that can't be broken up. So you would start there and simplify this expression to $|6| \div 3$. Then you can simplify the absolute value, because $|6| = 6$. Now the expression is $6 \div 3$, which you know is equal to 2.

Absolute Value with Variables

Things always get a little trickier when variables are involved. The big thing you have to remember when it comes to variables is that while the absolute value will always be positive, the number inside the absolute value bars could be positive or negative. Say you're told that $|6| = x$. That's easy enough, because you can simplify $|6|$ to 6 to find that $6 = x$. But what if you're told $|x| = 6$? Now things get a little more complicated.

Notice that it's impossible to be given an equation such as $|x| = -6$. An absolute value is positive, so there's no way the absolute value of anything can equal a negative number.

You know that the absolute value of *x* is 6. In other words, *x* has a magnitude of 6, or it is six units away from zero. What number is that true about? It's certainly true about 6. But it's also true about –6. Check it out: you can plug either 6 or –6 in for *x*, and this equation will still be true. It's true that $|6| = 6$, and it's true that $|-6| = 6$. So this equation actually has two possible solutions: $x = 6$ or $x = -6$.

Absolute value equations with a variable inside the absolute value symbol can have two solutions. This is similar to exponent equations: equations with a variable raised to an even exponent can have two solutions as well. When you are told $x^2 = 9$, it could be true that $x = 3$, but it could also be true that $x = -3$. You can check for yourself: whether you plug in 3 or –3 for *x*, x^2 will still equal 9. Now take a look at the equation $x^2 = 9$ and imagine that you're going to try to solve for *x*. Usually when you want to undo a square, you use a square root. Take the square root of both sides of the equation and rewrite it as $\sqrt{x^2} = \sqrt{9}$. You know that $\sqrt{9} = 3$, so you can rewrite the equation as $\sqrt{x^2} = 3$. It has already been determined that both 3 and –3 are possible values for *x*, but how do you get there algebraically? Anytime you take the root of a square (like you are doing here), you're going to have both a positive and a negative solution. In other words, you can either think of or rewrite $\sqrt{x^2}$ as $|x|$.

$\sqrt{x^2}$ and $|x|$ mean exactly the same thing; they are perfect replacements for one another. In the first, you are taking a number, squaring it, and then rooting it, which puts you right back where you started, except the result will always be positive. In the second, you are taking a number and either keeping it as positive or changing it to be positive.

When you determine $|x|=3$, you have two possibilities. One is that $x=3$ and the other is that $x=-3$.

Solving Absolute Value Equations

This section will take all of the information you've learned so far about absolute value equations and put it together so you have a set of steps to follow when you encounter one of these problems. Consider the equation $\frac{|x+3|-1}{2}=7$. Here are the steps:

1. **Treat the absolute value symbol like parentheses and isolate it to one side of the equation.** In this equation, the first step you have to take toward simplification is multiplying both sides of the equation by 2. Now you have $|x+3|-1=14$. You should move the 1 over to the right side of the equation, which you do by adding 1 to both sides. Now you have $|x+3|=15$.

Remember, even though you generally do PEMDAS in reverse to simplify an equation, you must treat expressions that are the numerator or denominator of a fraction as if they are in parentheses.

2. **Create two possible equations.** You need to rewrite this equation as two equations without absolute value signs: one where the expression

inside the absolute value equals the positive solution, and the other where the expression inside the absolute value equals the negative solution. Given the equation $|x+3|=15$, it's certainly possible that what's inside the absolute value sign is positive, in which case $x+3=15$. But, the expression inside the absolute value sign could also be negative! If that's the case, $x+3=-15$.

3. **Solve both equations separately.** It's important to remember that these equations cannot both be true at the same time, so this isn't a system of equations you solve together. Instead, these are two possible equations that will give the two possible solutions. Your possible equations are $x+3=15$ or $x+3=-15$. When you solve these, you'll find that it's possible that $x=12$ or $x=-18$.
4. **Double-check your work by plugging both of your answers into the original equation.** You have decided it's possible that $x=12$ or $x=-18$, so these are the two values of *x* that should complete your absolute value equation truthfully. The original equation was $\frac{|x+3|-1}{2}=7$. If you plug in $x=12$, you get $7=7$. If you plug in $x=-18$, you get $7=7$. So these are both correct!

You may or may not need to write two solutions as two separate equations, such as "$x = 12$ or $x = -18$." (These equations must be joined with the word *or*.) Instead, you may have to write a list of possible answers for x. In that case, you would write $x = 12, -18$.

Chapter 18 Exercises

Find all possible values of x for each of the following equations.

1. $|x| = 6$
2. $|6| = x$
3. $|6| = -x$
4. $|-x| = 6$
5. $|x| = 6 - 3$
6. $|3| = x - 2$
7. $|-3| + x = 2$
8. $|x + 1| = 4$
9. $|x - 1| = 9$
10. $|1 + x| = 6$
11. $|2 - x| = 5$
12. $|2x| = 4$
13. $|-2x| = 4$
14. $\left|\frac{x}{2}\right| = 6$
15. $\left|\frac{6}{x}\right| = 3$
16. $|2x + 1| = 9$
17. $|2x - 1| = 9$
18. $|2 + 2x| = 10$
19. $|2 - 2x| = 10$
20. $|x| = 0$

CHAPTER 19

Plane Figure Geometry

The name "plane figure geometry" sounds way scarier than it is. When you hear someone talk about figures in a plane, they're just talking about two-dimensional (2-D) shapes—shapes like circles, squares, and triangles that you can draw on a piece of paper. This chapter will give you an overview of the properties of shapes such as triangles, rectangles, and circles.

Two-Dimensional Shapes

In one dimension, you can only draw lines. Think of directions. Lines can only be measured in length; you can only measure them in one direction. But two-dimensional shapes have something in addition to length; they also have area. Area is an enclosed space that can be filled in. Area has two dimensions: length and width. It takes up space when you draw it on a piece of paper, but not in the real world. These are the shapes you deal with in plane figure geometry.

Solid figure geometry, by contrast, involves three-dimensional (3-D) figures. These are figures that can't be drawn on paper, because they take up space in the real world. Solid figures are figures like cubes, spheres, and pyramids. They have length, width, and height.

Two-dimensional shapes are flat. Everything about them is created by length and width; they don't have the "up" dimension that 3-D shapes have.

Polygons

A **polygon** is a plane figure with at least three straight sides. This means a polygon has to have sides that are straight lines, and those lines have to come together to create an enclosed space.

Here are some shapes that are polygons:

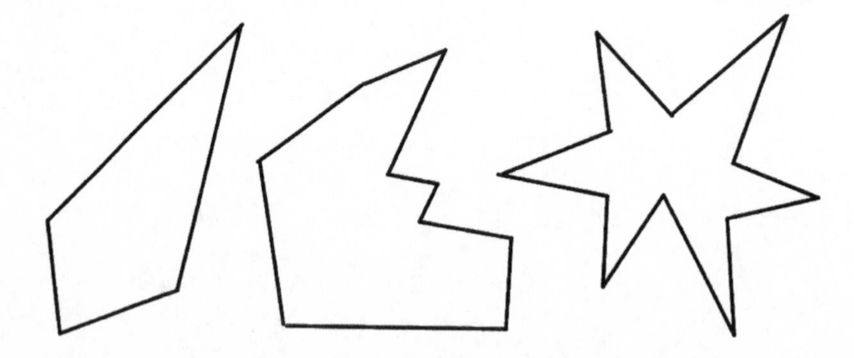

These shapes don't have any special names, but they are still polygons. Their sides are made of straight line segments, and they create an enclosed space. In other words, you could "color within the lines" and fill in each of these shapes.

What wouldn't be a polygon?

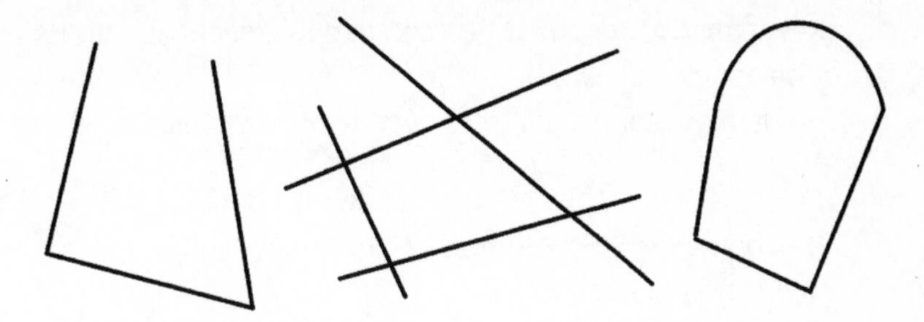

These images are not polygons. The first image is open on one side—you couldn't "color within the lines" here. Polygons have to create an enclosed space.

The second image is not a polygon because the line segments do not stop when they meet the other line segments. In other words, they don't create "corners" for the shape. While that second image does *contain* a polygon (the enclosed four-sided shape), the image itself is not a polygon.

A "corner" of a polygon is called a **vertex**. The sides of a polygon must come together at its vertices (that's the plural of *vertex*).

The third image is not a polygon because, while it is an enclosed shape, its sides are not composed of straight line segments.

Describing Polygons

So you know that a polygon is a flat, enclosed space whose edges are made out of straight lines, and that those straight lines meet at the vertices. That includes a *lot* of different shapes. And as you can see from the previous examples, a lot of them look pretty strange.

You have a few different ways you can describe polygons. Some of the most common ways you classify polygons are as concave or convex, simple or complex, and regular or irregular.

Concave Versus Convex

In convex polygons, the vertices point outward. Because of that, a line drawn through a convex polygon that intersects one boundary will intersect a total of exactly two boundaries.

Here are some examples of convex polygons:

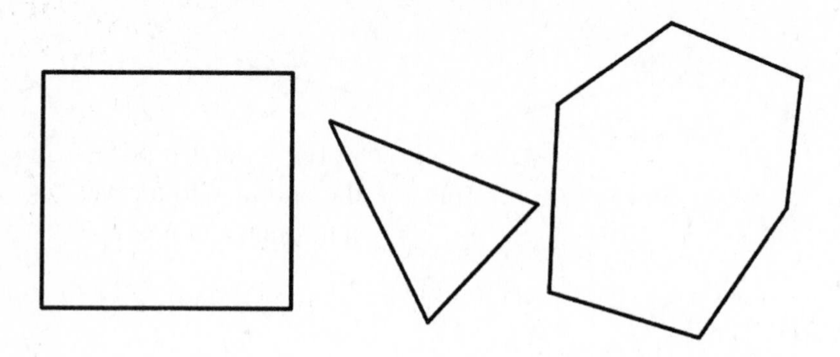

The vertices point outward, and if you were to draw a line through any of these shapes that intersected one of the edges, that line would intersect *exactly one* other edge. See?

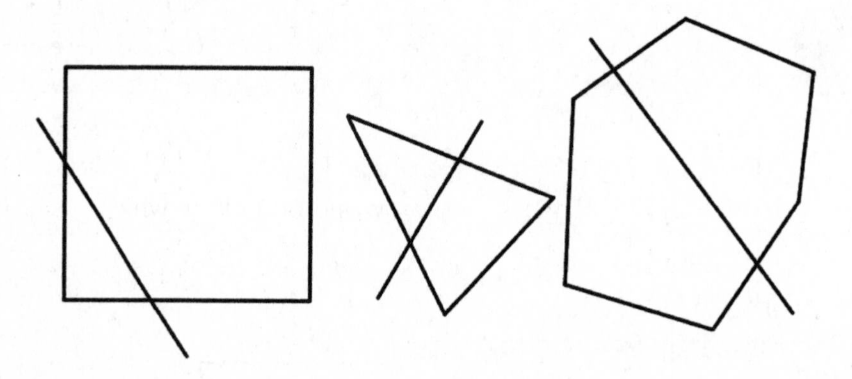

In a concave polygon, however, at least one of the vertices points inward. Because of this, if you were to draw a line that intersects one of the edges, you might intersect *more than one* other edge.

The following drawings are examples of concave polygons.

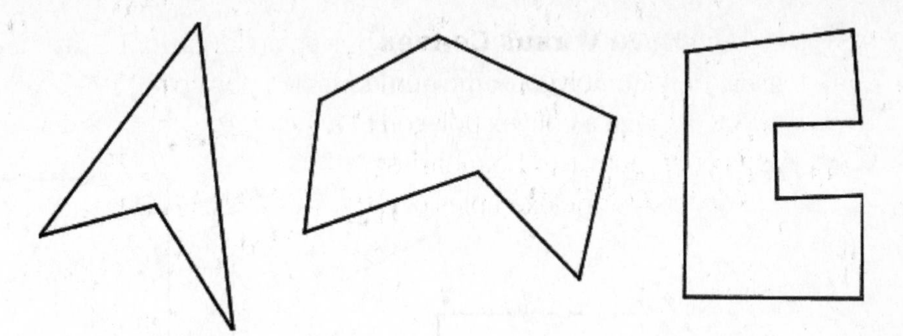

You might like to remember that con*cave* polygons are the ones that have some sort of indent, like the entrance to a cave. The following image shows how a line can be drawn through a concave polygon so that it intersects more than two sides.

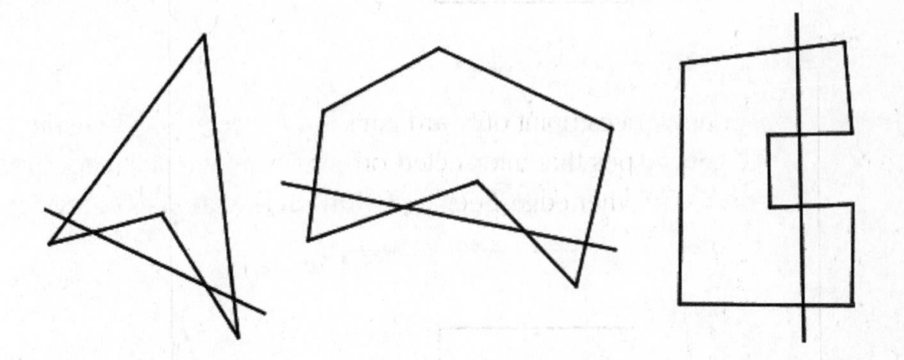

Simple Versus Complex

Simple polygons have only one boundary, which never crosses itself. These are some examples of simple polygons:

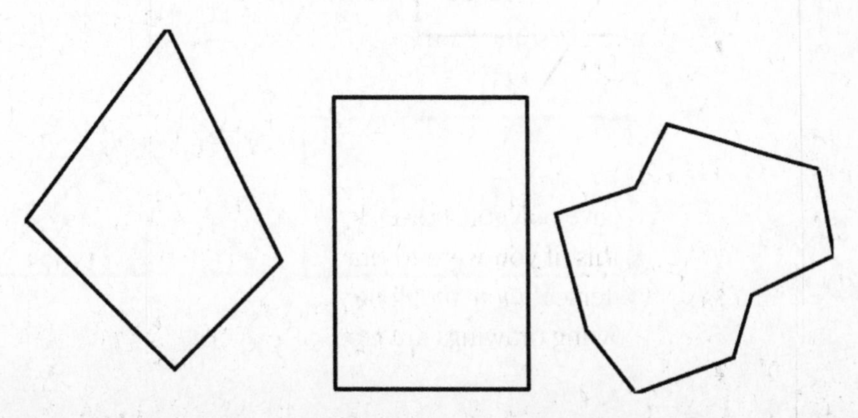

On the other hand, complex polygons have a boundary that intersects itself! Here are some examples of complex polygons:

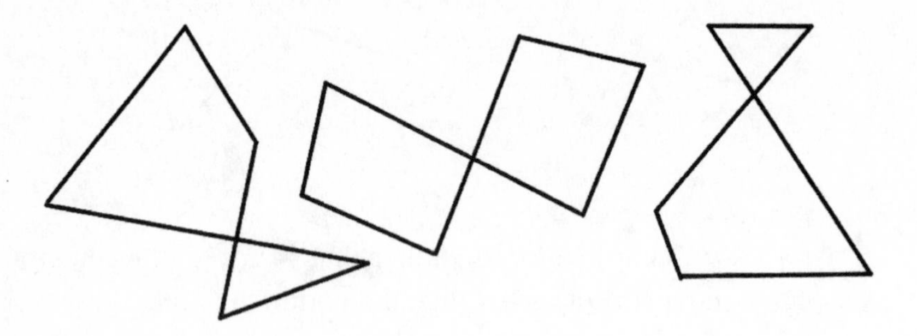

Regular Versus Irregular

Another way to classify polygons is as "regular" or "irregular." In a regular polygon, all angles are equal and all sides are equal. If it doesn't fit these criteria, the polygon is irregular. Here are some examples of regular polygons:

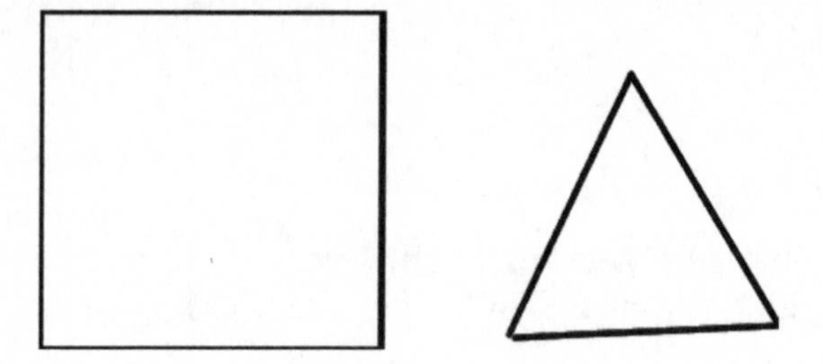

And here are some examples of irregular polygons:

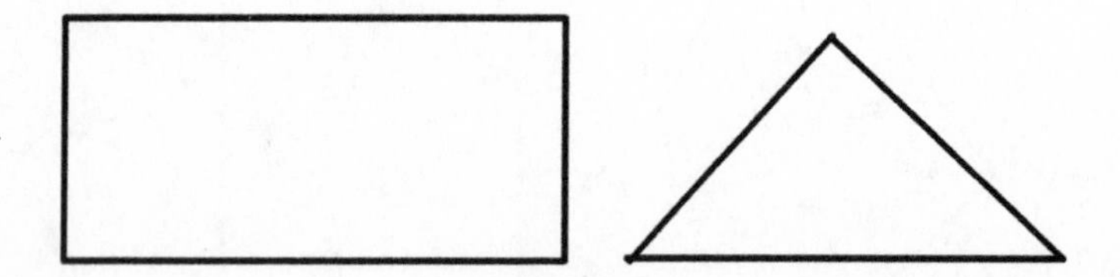

Measuring Angles

Measuring angles is fundamentally different from measuring something like length. When you measure length (or really almost anything you might measure), the size of the thing you're measuring is important. The length of an ant is going to be a lot shorter than the length of a subway car, which is going to be a lot less than the length of a blue whale.

Angles don't really work that way. Angles are kind of like the fractions of the geometry world. Remember that fractions are relative numbers: they just tell you how much you have out of the total number of pieces, but they don't tell you how big each piece is.

An **angle** is the space between two intersecting lines. Here are some examples of angles:

Angles aren't measured in inches, feet, or meters: they are measured in something called degrees. **Degrees** are the units of measurement for angles; they tell you how narrow or wide an angle is. In other words, the measure of an angle only tells you how slanted the lines that form the angle are in relationship to one another. It does *not* tell you how long those angles are. Take a look at these three angles:

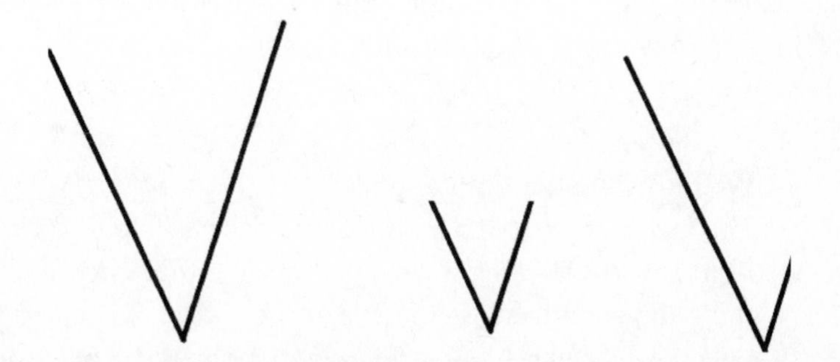

These angles all have the same degree measure. The pictures are different sizes, but you can see that the lines always come together with the same degree of slant toward one another. If you made a little wedge that fit perfectly into the first angle, you could still fit that wedge perfectly into the corner of the second angle and the corner of the third angle. So the degree measure doesn't tell you how big the two intersecting lines are. It tells you the relationship of the two lines to one another.

It's probably easiest to understand angle measurement by starting with the angle formed by two perpendicular lines. Perpendicular lines intersect in a way that makes a perfectly square corner when they meet. Take a look at some examples of perpendicular lines:

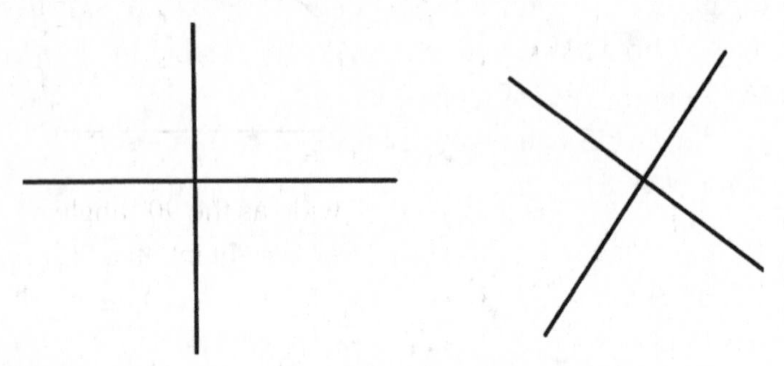

In these pictures, you can see that whatever the slant of the lines is, where they come together, they form a corner that looks like the corner of a square. The measure of that angle is 90°. That little floating circle is the symbol for the word *degrees*, so you would read the expression 90° as "ninety degrees."

You call an angle that measures 90° a **right angle**. Right angles are all around you. The wall of the room you're in meets the floor at a right angle. It also meets the ceiling at a right angle . . . otherwise, the ceiling would probably cave in. The corner of your math book is a right angle, because the corner of any rectangle is a right angle. You could look around the room you're in right now and probably count dozens of right angles.

Why is it called a right angle?
The fact that it's called a *right* angle has nothing to do with the idea of right and left. A right angle can open in either direction. It's called a right angle in the sense of "correct" or "true," meaning that if you build a wall against a floor at a right angle, it will be strong and hold up. There is also some sense of the direction of "right" or "correct" being up, like when you talk about walking "upright."

As an angle gets narrower, its degree measure goes down. If a right angle is 90°, then any angle narrower than a right angle must have a degree measure of less than 90°. The narrower the angle gets, the smaller the degree measure gets. Here's what a 45° angle looks like:

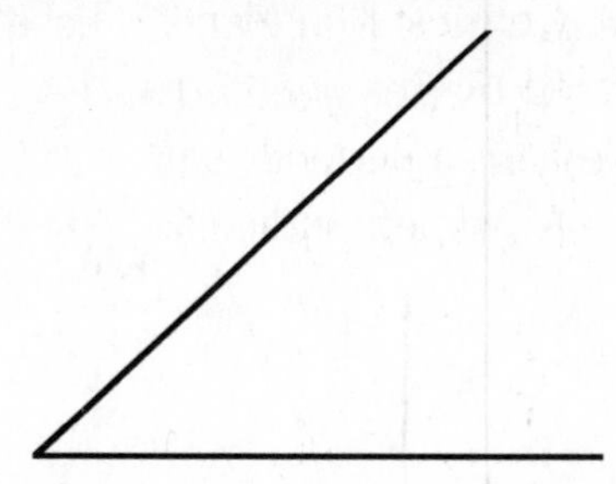

You can see that it's half as wide as the 90° angle. In fact, if you took two of the 45° angles, you could fit them right into the 90° angle from before, like this:

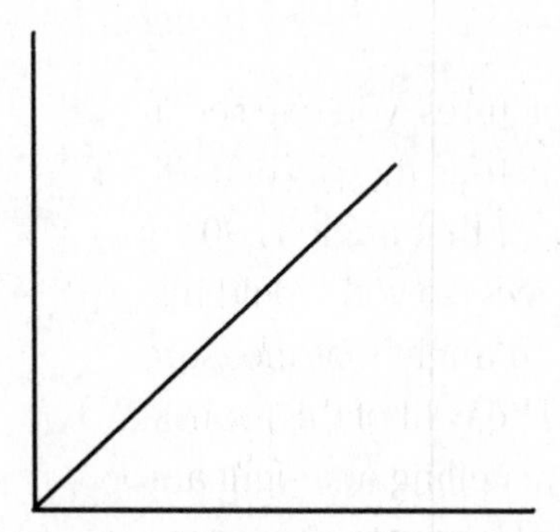

So the angle measure tells you how narrow or wide the angle is in comparison to other angles. Here are some other angles that are smaller than 90°:

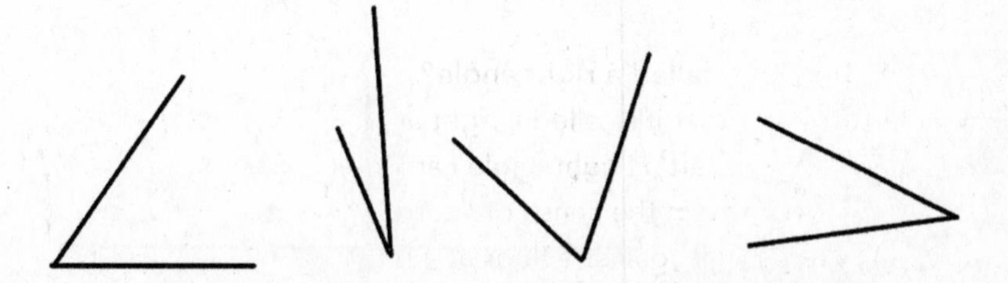

These are called **acute** angles. An acute angle is an angle that is smaller than 90°. You can remember this by thinking how "cute" these angles are, since they are small. Keep in mind that an angle measure can't be negative: geometry is based on shapes, and shapes exist in the real world. Since you don't have negative numbers in the real world, you don't have them in shapes. An angle measure also can't be zero: then it wouldn't be an angle, so you couldn't measure it. Acute angles have a measure between 0° and 90°.

What about angles bigger than 90°? An angle bigger than 90° might look something like this:

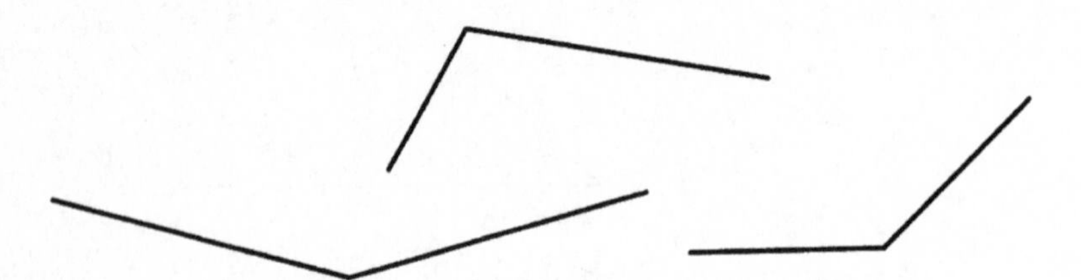

Angles bigger than 90° are called **obtuse** angles. Outside of math, *obtuse* is a word that is used to describe someone who's not very aware of what's going on. (That's a good SAT word, by the way.) You know how you use the word *sharp* sometimes to mean "smart" or "observant"? The opposite of sharp is dull, and that's exactly what *obtuse* means in math.

Adding Angles

Remember how you could take two 45° angles and fit them into a 90° angle? That's because angle degrees add and subtract just like you'd think they would. So what happens when you put two 90° angles next to one another and add them up? Two things happen. First, you end up with a straight line. Take a look:

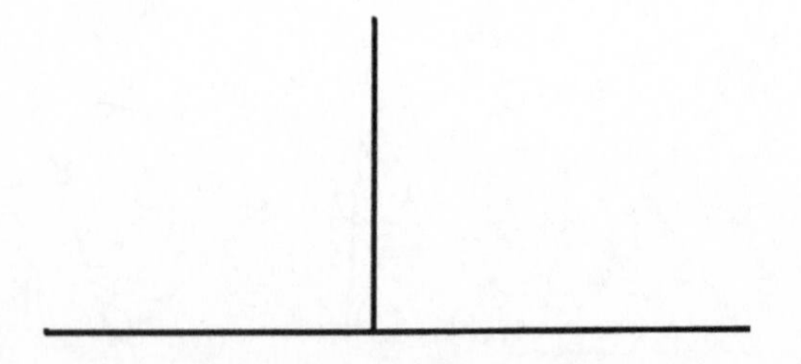

Second, you know that the total angle measure of that straight line must be 180°, because it's made from the sum of two 90° angles. So now you know that the degree measure of a straight line is always going to be 180°.

Sum of the Angles of a Polygon

Angles are a really important part of understanding and classifying shapes. One piece of information you can use to understand shapes is the sum of their angles. For example, you know that the sum of the angles in a triangle will always be 180°. It doesn't matter how you draw the triangle or what it looks like; if it's a triangle, the sum of its angles will be 180°.

A triangle, as you know, has three angles: that's why it's called a "tri-angle." If all those angles were the same, they would each be 60°, and the triangle would look like this:

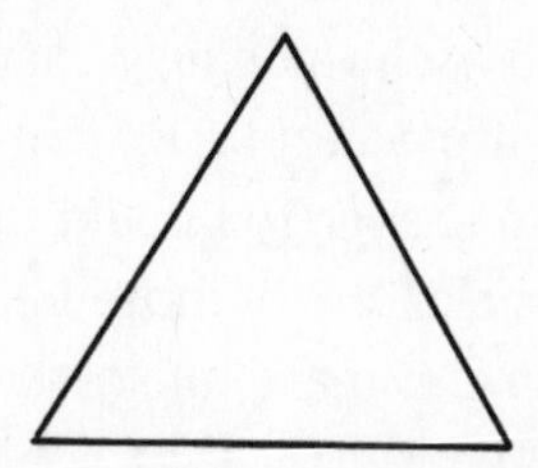

But what happens when you try to make one of those angles bigger, increasing its angle measure? Say you take the angle at the top of the picture and make it bigger. There's only so much angle to go around, so when you make the angle on the top bigger, the other angles must get smaller. Their sum stays at 180° even though the value of each angle might be changing.

Therefore, the sum of the angles in a polygon that has three sides (called a triangle) is 180°. What about the sum of the angles of a square? Well, you know that the corners of a square are all right angles, and each right angle measures 90°. If you have four corners, and they each measure 90°, the sum of the angles of a square must be $4 \times 90° = 360°$. The same thing is true for a rectangle, for the same reason: rectangles have four corners that are 90° each, so the sum of their angles must be $4 \times 90° = 360°$.

Why is the sum of the angles of a square or rectangle twice as big as the sum of the angles of a triangle? Because a square or rectangle can be drawn as two triangles. Check it out:

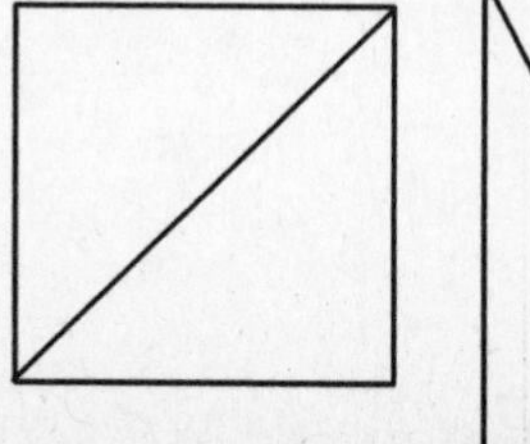

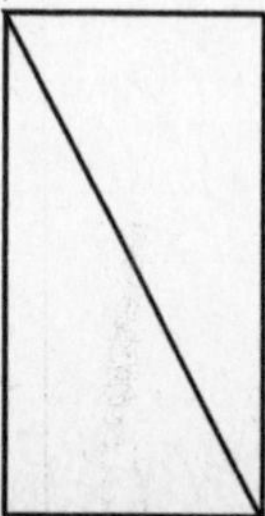

Since their angles are made out of the angles of two triangles, the sum of their angles has to be the sum of the angles of two triangles, or $180° \times 2 = 360°$. In fact, this rule isn't just true for rectangles and squares. It's true for any polygon that has four sides! Why? There are two ways to think about it. When you start with a rectangle, each of the four angles of that rectangle is equal to 90°. Much like you saw with a triangle, if you make any of those angles wider, another angle has to get narrower to compensate. There is only so much angle to go around: if you widen it in one place, you have to narrow it somewhere else.

If one angle gets wider, another has to get narrower to compensate. In the end, although each value may change, their sum remains the same. You can think about a piece of string that you are going to bend four times to make a four-sided shape. You could make every bend 90° and have a rectangle, or if you make one of those bends wider, somewhere along the line you're going to have to make one of the bends narrower to preserve the four sides.

Also keep in mind that any four-sided polygon can be made into two triangles. Check it out:

There are a lot of different ways to make a quadrilateral into two triangles. As long as you connect one vertex to the opposite vertex, you'll have two triangles.

When the number of sides in a polygon increases by one, the sum of the angles of that polygon increases by 180°. Why? Because you can fit another triangle into the polygon for every side you add. From this fact, you can derive a formula for the sum of the interior angles of any simple polygon.

The sum of the interior angles of a polygon with *n* sides is $180°(n-2)$. In other words, plug the number of sides in for *n*, and this formula will tell you the sum of the angles of the polygon.

Classifying Triangles and Quadrilaterals

Triangles are simple polygons that have three sides and three angles. **Quadrilaterals** are simple polygons that have four sides and four angles.

The prefix *quad–* means four, and the root *lateral* has to do with lines. So *quadrilateral* means "four sides," and you know that all four-sided shapes have four angles. *Quadrangle* is actually a word in math, but you won't use it as much as you use *quadrilateral*.

There are a lot of different ways to classify triangles, but for now there are three main types of triangles you should know by name.

Right Triangle

A right triangle is a triangle where one of the angles is 90°. That 90° angle will be the biggest angle in the triangle, because all three angles have to add up to 180°, so if one of the angles is 90°, the other two have to add up to 90° together.

The side of a right triangle that's across from the right angle is called the **hypotenuse**. It is the longest side of a right triangle. The other two sides, which come together to make the right angle, are called the **legs** of a right triangle.

Equilateral Triangle

An equilateral triangle has three sides of equal length. Because the sides are of equal length, the angles are also of equal measure: they're all 60°.

A triangle can be both a right triangle and an isosceles triangle! That's called an isosceles right triangle. The two sides that make the right angle will be the same length, and the angles across from them will each be 45°.

Isosceles Triangle

An isosceles triangle has two sides of equal length and two angles of equal measure. The two similar angles will be across from the two sides of the same length. The third, different side can be either the longest side or the shortest side.

The longest side of any triangle will be across from the biggest angle of that triangle. The shortest side of any triangle will be across from the smallest angle of that triangle. If two angles in a triangle are the same, the sides across from those angles are the same length.

When it comes to classifying quadrilaterals, it is easiest to define some different kinds of quadrilaterals and then see how those classifications overlap. Here are some types of quadrilaterals, which are shapes with four sides.

1. **Square.** A square has four sides of equal length and four angles of equal measure. All the angles are 90°.
2. **Rectangle.** A rectangle has four angles that measure 90°. Because the angles are all 90°, each side is parallel to (and the same length as) the side across from it.
3. **Parallelogram.** A parallelogram has two sets of parallel sides. (Maybe you can guess where the name comes from.)
4. **Trapezoid.** A trapezoid has at least one pair of parallel sides.
5. **Rhombus.** A rhombus is a special kind of parallelogram: it has four sides of equal length, and the opposite sides are parallel to one another.

Some textbooks define a trapezoid as having *exactly* one pair of parallel sides. That definition stops working in very advanced math, but it works fine for now!

Other Quadrilaterals

There are other irregular quadrilaterals. Some have no name; they can only be defined as having four sides and four angles. Others are named: a kite, for example, is made by a pair of two sides of the same length on one side of the quadrilateral and a pair of two sides of a different length on the other side.

Here are the most typical versions of each shape.

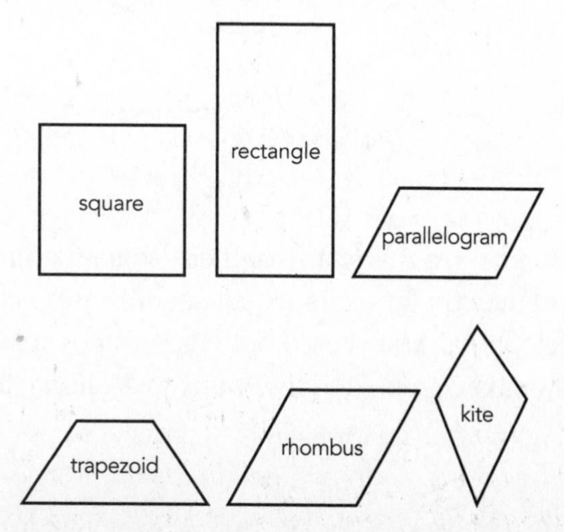

Now, these groups overlap quite a bit. For example, think about a square. Sure, it's a square, but it has four 90° angles, so it's also a rectangle. And it has two sets of parallel sides, so it's also a parallelogram. And it has four equal sides, so it's also a rhombus. And by definition, it's technically a trapezoid too, because it has at least one pair of equal sides. It can help to think about these types of shapes in terms of which categories are completely included in other categories, and which ones may overlap.

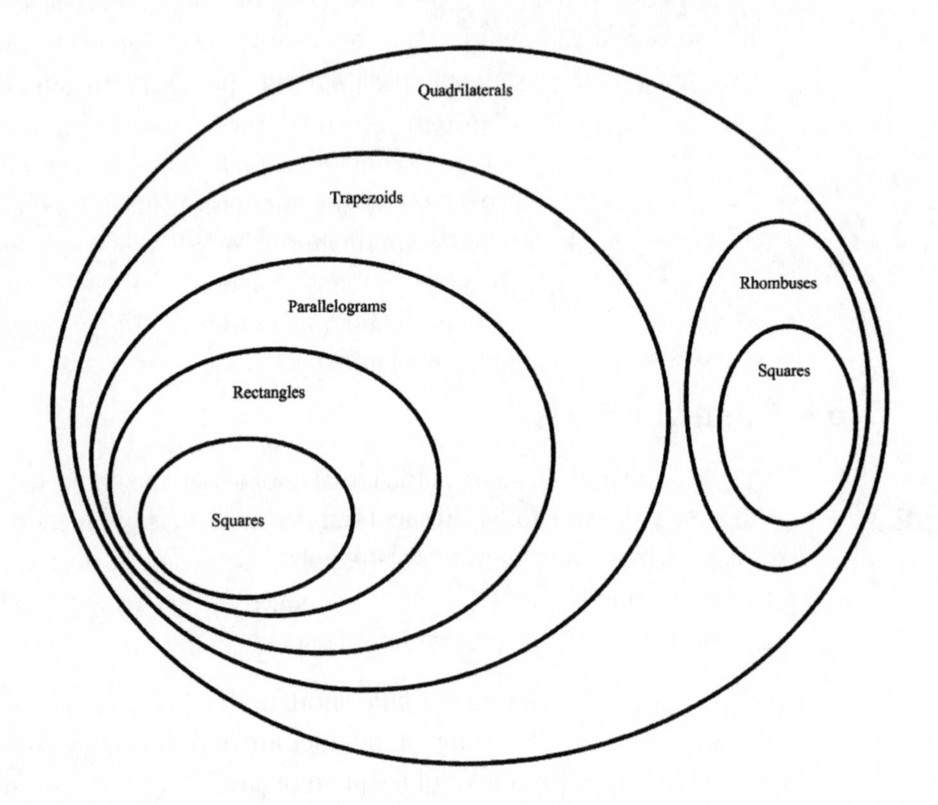

You can see that many of these shape definitions are "layered" within one another. So for example, all squares are rectangles. All rectangles are parallelograms. You should note that it's possible for a shape to be in both the "rhombus" group and the "trapezoid" group, but only if it's a square.

It's best to use the most descriptive name for a shape. If you know something's a square, you should call it a square. It's more specific and helpful than calling it a parallelogram, even though it is technically also a parallelogram.

Perimeter Equations

The **perimeter** of a polygon is the total measure of the length of the sides of polygon. *Peri* means "around," so you're measuring the distance *around* a shape. You can think of the perimeter as a fence: how much fence would you need to build a border around the shape? And just like you measure a fence in length, you measure perimeter in length. When you add up the lengths of all the sides, you'll have the perimeter of the shape. That's pretty much the whole formula for perimeter: the perimeter of any polygon can be found by adding the lengths of all its sides.

Make sure your information is all in the same units before you start adding! You can't add inches and feet, for example, until you make them both into inches or both into feet.

Sometimes you can take a little shortcut, because you know from the definition of a shape that some of the sides are equal in length. For example, the formula for the perimeter of a square or a rhombus is 4*s*, where *s* stands for the length of a side. That's because every side of a square or rhombus is the same, so when you add up the sides, you get $s+s+s+s$, which you know can be simplified to 4*s*. Even though this formula involves multiplication, it really comes from addition. In a rectangle where the longer side is called the length (l) and the shorter side is called the width (w), you know that the perimeter will be $l+l+w+w$, which you can rewrite as $2l+2w$ or $2(l+w)$. These are the same because of the distributive property, so you can use whichever one you like best.

Area Equations

If the perimeter is the "fence" around a shape, the area is the "paint" it would take to cover the surface of that shape. While you measure perimeter in lengths such as feet or inches, you measure area in units such as square feet or square inches.

Whatever the unit of measure of length is, the unit of measure of area will be that unit squared. For example, if you're measuring the perimeter in centimeters (*cm*), you will measure the area in square centimeters (cm^2).

To find the area of a rectangle, you must multiply the width by the length. So the formula for the area of a rectangle is area $A = lw$, where A stands for the area, l stands for the length and w stands for the width. Say you have a rectangle with a length of 4 inches and a width of 6 inches. The perimeter shows that 20 inches of fence surround the shape, but the area is 24 square inches; you could fit 24 square tiles on that rectangle.

Since a square is a rectangle, the same formula applies: the area of a square is the length times the width. But wait! Since it's a square, the length and width are the same. So you could just say that, given a square where each side has a length of s, the area of that square would be $s \times s$, or s^2.

Why is it called "squared" when you raise something to the second power?
Because when you raise something to the second power, you multiply that number by itself, which is the same thing you do when you find the area of a *square*.

When you find the perimeter of a triangle, you just add up the lengths of the sides. That's how you find the perimeter of any polygon. But what about the area of a triangle? Remember how any four-sided shape can be made into two triangles? A triangle is really just half of a rectangle. The area of a triangle is equal to one half of the base times the height. The formula is usually written as $a = \frac{1}{2}bh$, where a stands for area, b stands for the length of the base, and h stands for the height.

Just like you only measure your height when you are standing up straight, the height of a triangle can only be measured by a side that is perpendicular to the base. In a right triangle, this is simple, because the two sides of the triangle that form the right angle are perpendicular to one another. In any other triangle, you must find or be told the height of the triangle, based on a line that is perpendicular to the base.

The areas of quadrilaterals that are not squares or rectangles can be memorized as formulas, but you can always find the area or rediscover the formula by dividing a quadrilateral into squares, rectangles, and triangles. Here are the formulas for some other common quadrilaterals.

1. **Area of a rhombus:** The area of a rhombus is the base times the height, or $a = bh$. The base will be the length of one of the sides, and the height (like the height of a triangle) will be measured as a line that is perpendicular to the base.
2. **Area of trapezoid:** In a trapezoid, the two parallel sides are called the bases. The area of a trapezoid is the average of the bases, times the height. In other words, $a = (\frac{b_1 + b_2}{2})h$. You add the two bases and divide by 2, and then multiply that number by the height.
3. **Area of a parallelogram:** The area of a parallelogram is the base times the height ($a = bh$). Any side you want can be the base; the height is measured from a line that is perpendicular to whichever side is the base.

Remember that when "height" is included in a formula, it has to be measured perpendicular to the base. The "height" is not necessarily the length of one of the sides; that is only true for squares and rectangles. Just as when you are measured to see how tall you are you stand up straight, when you measure the height of any shape, you have to find the distance, measured perpendicular to the base, from the base to the tallest point of the shape.

Chapter 19 Exercises

Answer the following questions.

1. How many interior angles does a triangle have?
2. How many interior angles does a quadrilateral have?
3. What is the sum of the angles of a triangle, in degrees?
4. What is the sum of the angles of a square, in degrees?
5. What is the sum of the angles of a rectangle, in degrees?
6. What is the sum of the angles of a six-sided polygon (called a hexagon), in degrees?
7. If one angle of a triangle measures 20° and one angle of a triangle measures 100°, what is the measure of the third angle?
8. If one angle of a triangle measures 45° and one angle of a triangle measures 50°, what is the measure of the third angle?
9. If one angle of a triangle measures 45° and another angle of the triangle also measures 45°, what is the measure of the third angle?
10. What is the name for a triangle that has two sides of equal length?
11. What is the name for a triangle that has one angle measuring 90°?
12. What is the name for a triangle where each angle measures 60°?
13. What is the name for a triangle with one angle measuring 90° and two angles measuring 45°?
14. What is the perimeter of a rectangle with a length of 5 inches and a width of 3 inches?
15. What is the area of a rectangle with a length of 5 inches and a width of 3 inches?
16. What is the perimeter of a square whose sides are 3 inches long?
17. What is the area of a square whose sides are 3 inches long?
18. What is the angle measure, in degrees, of a straight line?
19. What is the area of a triangle with a base of 3 inches and a height of 6 inches?
20. Is an 80° angle acute or obtuse?

CHAPTER 20

Test-Taking Strategies

There are many people who have no problem learning math, but still don't perform as well as they want to on tests. Some people just find that tests don't allow them to show what they're really capable of, whether it's because of test anxiety or for another reason. If you feel that way, you should know that you aren't alone. Unfortunately, the reality is that there needs to be some way to assess what you're learning and what you know, and most of the time this is done with tests. Because of this, it makes sense to try to get better at taking tests, because you want to be able to show off everything you know.

Understanding that tests aren't a good indicator of your abilities is a great first step in conquering testing problems. Good test scores aren't essential to understanding math, but they *are* essential to succeeding in math *class*. You *can* get better at taking tests!

Pacing and Timing

One of the things that sometimes causes stress during a test is being under time pressure. If you find that managing your time is really hard for you, you might talk to your teacher and see if you can have a little bit of extra time to take an exam. He or she might say no, but it never hurts to ask, and it may help your teacher to understand what you're going through.

Even if you can't get extra time, you can do a lot of things to help yourself deal with time pressure during a test. One thing that can help is to look for questions you know how to do, and do those questions first. It seems silly, but once you get a few questions right, it can help you relax and be better able to focus on the harder questions. You don't have to go in order! This is also a good idea because if you run out of time and have a few questions left unanswered at the end of the test, at least they won't be ones you definitely knew how to do.

Practice!

One thing that makes time pressure so difficult for many people to deal with is that it is a new experience, but there's no reason it has to be. You can practice solving problems under time constraints by creating a test environment for yourself at home. Try doing a few homework problems while being timed to get a feel for how long it takes, or set a goal and see how many homework problems you can complete in ten minutes. This will help you see that having a clock running doesn't really change anything. You can do this by trying to do a set of homework problems under a time limit. Of course, if you rush through any problems or don't finish, you'll want to keep working past the time limit, but it is helpful to get used to the feeling of having a time pressure.

Above all else, don't hurry and make silly mistakes when you know better. There's no reason to rush through a test, but there is a good reason not to: when you rush, you make mistakes. You can move along at a focused pace without rushing. It takes concentration, but luckily that's something you can practice. Try doing your homework while sitting in a quiet place with no distractions and work straight from the first problem to the last problem without taking any breaks. If that's hard, you might have to practice concentrating for shorter periods of time first. But if you can get good at concentrating on one thing for an hour or so at a time, your timing problem on tests should be greatly reduced.

Skills like multiplication, division, and factoring really benefit from drill practice, and even five minutes a day will make a difference. Improving these skills will not only save you time, but also help you avoid careless mistakes in your test answers!

How to Stay Relaxed

When it comes to learning math, what matters is the overall pattern of effort, improvement, and consistent hard work, not the grade you get on a test. So don't freak out! In order to do your best work, you will need to learn how to relax when being tested.

Of course, the best way to be calmer and more focused and to avoid panicking during a test is to be prepared. If you've studied the material and know how to do the problems, you won't feel as anxious. But sometimes you study and feel like you're ready, and then you open the test and there are things you don't understand. Or sometimes you study, but you feel so much pressure that you can't get your head clear.

When that happens, you should try to close your eyes for a few seconds and take a deep breath. When you are done, take a look at the test and find something you know how to do. Do that first.

Starting with the problems that you understand has four big benefits. First, it builds confidence and helps you remember how to apply what you've learned on the test. Second, it ensures you get to prove what you already know. Third, it lets you spend as much time as possible trying to understand the other questions without giving anything up. And finally, sometimes by doing the problems you understand, you'll learn or remember something that will help you with the remaining problems.

If you feel like you studied but don't understand the question, make sure you've read the directions and the question carefully. Ask for clarification from your teacher if you're allowed to do so. Try to think about what you've studied and see how it applies to your test. Even if you can't get a problem perfectly, do as much as you can. This will help your teacher to see where you're getting stuck.

Staying relaxed during the test is a little bit of a mind game. It's easiest if you've practiced your homework or studied in a similar condition. If you're always doing homework by looking at the answers or by looking things up along the way, you can't really be sure that you'll still be able to do the problems when that crutch is removed. You can help remove feelings of anxiety during a test by treating your homework like a test, and trying to work through each question on your own without looking for help or checking your answers repeatedly. It may make your homework take a little longer, but it saves you a lot of time during studying and a lot of anxiety on test day.

Looking for Clues

Sometimes you read a question on the test, and your mind just goes blank. It happens to everyone. If you miss one question, that's okay! Keep in mind that you can get unstuck by looking for clues in the question. You have an advantage when you take a math test: you know what it's testing. Think about what you've studied, and try to look for words or equations in the question that match what you practiced at home. This might help your brain to pick up on a clue that tells you which skills you should use for that particular problem.

The same is true for final exams, which can cause a lot of stress. When you look at a question, try to think about what type of question it is first. For example, what chapter in the book did it come from? What is the topic it's testing? This categorization allows your brain to pull up the information it knows about that particular topic, instead of trying to search through everything it knows to answer that one question. Usually if you categorize the question first, you have a better chance of remembering what you're supposed to do to solve that type of question.

One great way to get unstuck on a problem you don't quite understand is to start by simply taking one step that you *do* understand. Maybe you can start by simplifying the expression, or you notice that a number can be factored out of an expression. Look for something you *do* know, and it will often transform the problem into something you've seen before and know how to tackle.

Making a Plan

Once you know what kind of question you're dealing with, it's a good idea to stop and make sure you have a plan for where you're going. This might only take one or two seconds, but it can be helpful to think of "making a plan" as its own step. This way, you ensure that you understand the question and what is being asked of you before you start doing a lot of work in the wrong direction.

Maybe you notice the question has fractions that have to be added, so then the first step of the plan would probably be to find a common denominator. Maybe you notice that the equation you're being asked to solve has an x on both sides, so the first step might be to combine like terms. Maybe you can see that the equation involves addition, division, parentheses, and exponents, so you know you're going to have to follow PEMDAS and be very careful, simplifying one step at a time.

Thinking about questions in terms of a plan can help reduce anxiety. It also helps you take what you learned in class and on your homework and recognize when you should apply it to a test.

When you are doing your classwork or homework, don't just follow the steps you've learned without giving any thought to what kind of question you are solving. Otherwise, when the test comes, you won't know which skill to use where. Instead, try to start by clarifying to yourself what type of problems you're practicing and why the method of solution works.

Watching Your Work

Silly mistakes can be really frustrating, because you know that you didn't really get to show your stuff. Mistakes that you understand are the most fixable. When you make a silly mistake, like subtracting a number incorrectly or forgetting a negative sign, don't ignore it and assume it will go right next time. Instead, practice re-doing the problem correctly this time, to cement the correct pattern into your brain. That's how you fix it! Mistakes that are easy for you to fix are definitely mistakes you want to take the time to correct.

In math, it's important to be careful. Write clearly. Do your work one step at a time. When you have sloppy writing and try to do twenty steps at once in your head, mistakes happen. You know that it sometimes feels like a waste of time and energy to do everything out one step at a time, but if you're losing points and getting wrong answers because of silly mistakes, slowing down may be the way to fix it.

If you aren't 100% sure you can do something in your head, write it down. If your teacher asks to see your work, write it down. Make sure you give yourself enough space so that you can write methodically and clearly. It actually matters! If you can't see and follow the steps you're taking, you won't be able to follow them where they're supposed to lead you.

If you have time, you can and should double-check your work. Check through your arithmetic and make sure it's correct. Make sure you've completed every problem according to the directions. In algebra, when you've solved for a variable, you can check your work by plugging your solution into the initial equation and ensuring that it comes out to be a true equation. If it does, you know your answer is correct! If not, you need to re-check your work.

If you plug your solution back in for a variable to double-check your work and find out that your answer is wrong, what do you do now? It's really hard to find your own mistake by looking over what you've done. Instead, it's often easier to start by re-doing the problem carefully, step by step, as a second try at the question. You're more likely to re-solve a problem correctly than you are to catch the mistake you made when reviewing your work.

Don't Stop There

Math builds on itself. In English class, if you don't understand a story or a poem, you can leave it behind once the quiz ends and start fresh on the next test. In history class, if you just don't understand one chapter, you can leave it and then try to understand the next chapter, at least to some extent.

But math doesn't work that way. It's more like learning a language, or learning to play an instrument. Each chapter builds on the one before. So if you're lost, and the test doesn't go well, you shouldn't just forget it and move on. That's going to come back to bite you!

Instead, use the test as it's meant to be used: to assess what you know. If it doesn't go well, that's a sign that you aren't finished. Take the test home and try to work through the problems on your own. If you're lost, look for some help. You can ask a classmate, meet with your teacher, or even find lots of helpful videos and explanations online. Even an hour spent reviewing the test to figure out what went wrong could save you a lot of time and worry in the future.

With a good foundation, you'll be able to keep building on what you know about math. Without that good foundation, it will all fall apart. You've already taken a big step toward understanding math by taking the time to read this book. Learning math can be really tough. It really is its own language, and not everyone learns at the same pace, but like a language, it *can* be learned. You just have to stick with it, address the problems as they come up, and then build your knowledge one new piece at a time.

In the end, don't let math scare you! Unlike some other subjects in school, there's no place for disagreement. Math is a place where anyone and everyone can be equal. Math rewards you for taking the time to understand the information presented and to learn the rules of presenting it. Math is the great equalizer; it rewards you for your work. And you *can* understand it and master it if you're willing to ask for help when you get lost. So when you're preparing for your next math exam, think of it as a fair playing field—a chance to prove everything you've learned.

APPENDIX A

Solutions

Chapter 2

Integers: 0 2 –4 100 –70 12

Indicate whether each of the following numbers is a positive integer, a negative integer, a positive non-integer, or a negative non-integer.

1. Negative integer
2. Positive integer
3. Negative non-integer
4. Positive integer
5. Positive non-integer
6. Negative integer
7. Positive integer
8. Positive non-integer
9. Negative non-integer

Chapter 3

Divisibility Questions:

1. Yes
2. No
3. No
4. Yes
5. Yes
6. No
7. No

Multiples Questions:

1. Yes
2. No
3. Yes
4. No
5. No

Factoring Questions:

1. Yes
2. Yes
3. Yes
4. No
5. No

Positive Factor Questions:

1. 1
2. 1, 2, 3, 6
3. 1, 3, 7, 21
4. 1, 17
5. 1, 2, 3, 4, 6, 8, 9, 12, 16, 18, 24, 36, 48, 72, 144

Chapter 4

Prime or Not Prime:

1. Prime
2. Prime
3. Not Prime
4. Not Prime
5. Not Prime
6. Prime
7. Prime
8. Not Prime

Prime Notation:

1. 2
2. $7^1 2^1$
3. $2^2 3^2$
4. $2^1 3^2$
5. 2^6

Greatest Common Factor:

1. 7
2. 50
3. 1
4. 13
5. 18
6. 5
7. 22

Least Common Multiple:

1. 26
2. 100
3. 72
4. 63
5. 36
6. 48

Chapter 5

Simplify the following expressions.

1. $\frac{7}{6}$
2. $\frac{5}{9}$
3. $\frac{2}{5}$
4. 0
5. 1
6. $\frac{3}{8}$
7. 1
8. $\frac{1}{2}$

Convert the following fractions to mixed numbers.

1. 3
2. $3\frac{1}{4}$
3. $3\frac{1}{2}$

Convert the following mixed numbers to fractions.

1. $\frac{10}{3}$
2. $\frac{15}{2}$
3. $\frac{11}{9}$

Chapter 6

Convert each fraction to a decimal. If necessary, round to two decimal places.

1. .33
2. .5
3. .4
4. .4
5. .25
6. 4

Convert each fraction to a percent. If necessary, round to the nearest percent.

1. 75%
2. 80%
3. 133%
4. 150%
5. 20%
6. 57%

Convert each percent to a decimal. Leave off any unnecessary zeroes.

1. .5
2. .13
3. 1.22
4. .04

Convert each decimal to a percent.

1. 450%
2. 6%
3. 23%
4. 10%

Convert each decimal to a fraction. Write your answer in the most simplified form.

1. $\frac{1}{5}$
2. $\frac{3}{2}$
3. $\frac{3}{50}$
4. $\frac{6}{25}$
5. $\frac{3}{25}$
6. $\frac{21}{20}$

Convert each percent to a fraction. Write your answer in the most simplified form.

1. $\frac{2}{5}$
2. $\frac{3}{20}$
3. $\frac{1}{50}$
4. $\frac{7}{5}$
5. $\frac{3}{10}$
6. $\frac{1}{500}$

Chapter 7

Answer the following questions with one word or a short phrase.

1. Negative
2. Positive
3. Yes
4. Yes
5. Positive
6. Positive
7. Negative
8. Zero
9. Impossible or undefined
10. The original number
11. The original number
12. Yes
13. Yes
14. Yes
15. No

Chapter 8

Your answers may vary slightly, but they should start with the same integer and be within a few tenths of the estimates below.

1. 1.7
2. 7.1
3. 9.9
4. 5.5
5. 14.1
6. 4.5

Rewrite the following expressions so that each is expressed as an integer.

1. 8
2. 9
3. 1
4. 1
5. 9
6. 1
7. 8
8. 0
9. 75
10. 20
11. 0
12. 9
13. 0
14. 16
15. 16
16. 18

Chapter 9

Simplify the following expressions. Write each answer as an exponential expression.

1. 3^6
2. 2^1
3. 5^{10}
4. 7^2
5. 11^{11}
6. 3^5

Simplify the following expressions. Write your answer as an exponential expression.

1. 1^1
2. 2^1

3. 5^8
4. 7^2
5. 11^3
6. 3^3

Simplify the following expressions. Write your answer as an exponential expression.

1. 4
2. 10
3. $10\sqrt{10}$
4. 4
5. 2
6. 1
7. 3
8. 10

Simplify the following radicals as much as possible.

1. 10
2. $2\sqrt{5}$
3. $3\sqrt{5}$
4. $\sqrt{2}$
5. $\sqrt{1}$
6. $4\sqrt{5}$
7. $8\sqrt{2}$
8. $5\sqrt{3}$

Rewrite the following expressions in prime notation.

1. 2^4
2. 2^{12}
3. 5^{18}
4. $2^{10}3^5$
5. $2^6 5^6$
6. $2^{12}3^{12}$

Chapter 10

Rounding questions.

1. 35
2. 111.1
3. 100
4. 123.1231
5. 50
6. 4,329.3
7. 94
8. 4,000
9. 36
10. 41.2
11. 41.19

Simplify the following and express as one term in decimal form.

1. 4,560
2. 12
3. 4
4. .0025
5. 100
6. 400
7. 1,340
8. 1,200
9. .03407
10. .23

Chapter 11

Simplify the following expressions by following the order of operations. Your answer should be in the form of one number.

1. 12
2. 2
3. –13

4. 42.5
5. 19

Indicate whether each of the following expressions or equations is an example of the associative property, the commutative property, the identity property, or the distributive property.

1. Commutative property
2. Identity property
3. Associative property
4. Identity property
5. Commutative property
6. Associative property
7. Identity property
8. Associative property
9. Commutative property
10. Distributive property
11. Identity property
12. Commutative property
13. Distributive property

Chapter 12

Indicate how many terms there are in each of the following expressions.

1. Two
2. Four
3. One
4. One
5. Two
6. Three
7. One
8. Two
9. Two
10. One

Each of the following expressions has two like terms. Identify the two like terms.

1. $4x$ and $3x$
2. 2 and 9
3. $2y$ and $4y$
4. x^2 and $2x^2$
5. x^2 and $2x^2$
6. x and x
7. $\frac{3+4}{2}$ and 9

Chapter 13

Simplify the following expressions to the furthest extent possible.

1. $7x$
2. $11y$
3. $x+2y$
4. $-3x$
5. $4a$
6. 0
7. $b-a$
8. $2x+1$
9. $-2x$
10. $3y$
11. $6y$
12. x^2
13. 1
14. 2
15. 2
16. $2x$
17. $10y$
18. xyz
19. $6z^3$
20. $\frac{x}{5}$
21. $x+\frac{1}{2}$
22. $\frac{2+a}{3}$
23. $\frac{b}{6}$
24. $\frac{2}{x}$
25. x

26. $\frac{b}{4}$

27. $\frac{2}{s}+\frac{s}{2}$

28. $-\frac{a}{2}$

29. $\frac{2x}{y}$

30. 3

31. 1

32. $\frac{12}{y^2}$

33. $\frac{x^2}{9}$

34. x

35. $\frac{4}{3}y$ (could also be written as $\frac{4y}{3}$)

36. $2x$

37. $\frac{9}{y}$

38. 3

39. $-k$

40. $\frac{4}{5}$

41. x^3

42. $2y^2$

43. y

44. $d\sqrt{d}$

45. 2

46. x^6

47. b^2

48. $2x^3$

49. 0

50. y

Chapter 14

Isolate the variable in the following equations. Do not combine or simplify any of the numbers or solve for the variable—just do the first step.

1. $x = 10 + 2$
2. $y = -8 - 3$
3. $b = 0 - 4$
4. $a = 3 + 3$
5. $a = \frac{9}{3}$
6. $x = 7 \times 2$
7. $x = \frac{8}{8}$
8. $\sqrt{x^2} = \sqrt{25}$
9. $x = 4^2$
10. $x = \sqrt{13}$ or $x = -\sqrt{13}$

Solve the following equations.

1. $x = 10$
2. $x = 36$
3. $x = 5$
4. $x = 36$
5. $x = 9$ or $x = -9$ (This "or" does not mean that either one is an acceptable answer. It means that both answers are possible and that you can't actually finish solving to just one value.)
6. $x = 81$
7. $b = 12$
8. $b = 18$
9. $b = 5$
10. $b = 45$
11. $b = 1$ or $b = -1$
12. $x = 0$

Chapter 15

Decide whether each of the following illustrates the distributive, the commutative, the associative, or the identity property.

1. Distributive property
2. Identity property
3. Commutative property
4. Distributive property
5. Identity property
6. Commutative property
7. Associative property
8. Identity property

Solve the following equations for x.

1. $x = 0$
2. $x = 25$
3. $x = 5$
4. $x = \frac{4}{5}$
5. $x = 8$
6. $x = 2$
7. $x = 18$
8. $x = 34$
9. $x = -6$
10. $x = -2$
11. $x = -4$
12. $x = -\frac{1}{2}$
13. $x = 0$
14. $x = 4$
15. $x = 7$
16. $x = 20$
17. $x = 1$

18. $x = 3$
19. $x = 14$
20. $x = -7$

Chapter 16

Use substitution to solve for y in the following equations.

1. $y = 4$
2. $y = 13$
3. $y = 2$
4. $y = 2$
5. $y = 16$

Use substitution to solve for x and y in the following systems of equations.

1. $x = 5,\ y = 0$
2. $x = 15,\ y = 5$
3. $x = 2,\ y = 1$
4. $x = 5,\ y = 7$
5. $x = 1,\ y = 1$
6. $x = 4,\ y = -4$
7. $x = 3,\ y = 1$
8. $x = 1,\ y = 4$
9. $x = 2,\ y = 6$

Answer each of the following questions.

1. 18
2. 9
3. 1
4. 8
5. 13
6. –1

Use the formulas given to answer the questions below.

1. 500 bolts
2. 1,200 nails
3. 10 hours
4. $100

1. $20
2. $16
3. $20
4. $21

Chapter 17

Fill in each blank with either > or <.

1. $3 < 5$
2. $-3 > -5$
3. $-3 < 2$
4. $-10 > -100$
5. $1.5 > 1.4$
6. $-1.5 < -1.4$

Simplify the inequality.

1. $x < 13$
2. $y < 11$
3. $x > -17$
4. $y < -2$
5. $x < 2$
6. $x > 63$
7. $y > \frac{5}{2}$
8. $x < 28$
9. $y > -\frac{1}{3}$
10. $x > -1$
11. $b < 5$
12. $b < -5$
13. $b > 10$

Chapter 18

Find all possible values of x for each of the following equations.

1. $x = 6$ or -6
2. $x = 6$
3. $x = -6$
4. $x = 6$ or -6
5. $x = 3$ or -3
6. $x = 5$
7. $x = -1$
8. $x = 3$ or -5
9. $x = 10$ or -8
10. $x = 5$ or -7
11. $x = -3$ or 7
12. $x = 2$ or -2
13. $x = 2$ or -2
14. $x = 12$ or -12
15. $x = 2$ or -2
16. $x = 4$ or -5
17. $x = 5$ or -4
18. $x = 4$ or -6
19. $x = -4$ or 6
20. $x = 0$

Chapter 19

Answer the following questions.

1. 3
2. 4
3. 180°
4. 360°
5. 360°
6. 720°
7. 60°
8. 85°
9. 90°
10. Isosceles triangle
11. Right triangle
12. Equilateral triangle
13. Isosceles right triangle
14. 16 inches
15. 15 square inches (can also be written as 15 in^2)
16. 12 inches
17. 9 square inches (can also be written as 9 in^2)
18. 180°
19. 9 square inches (can also be written as 9 in^2)
20. Acute

APPENDIX B

Glossary

Absolute value

The distance of a number from zero on the number line. Absolute value is always positive.

Acute

An angle whose measure is less than 90°.

Angle

The shape formed by the meeting or intersection of two lines or line segments, usually measured in degrees.

Approximately

Imprecise but roughly or closely equal to.

Area

The size of a surface of a shape, usually measured in square units such as square inches or square feet.

Associative property

The property of numbers that says that when numbers are being multiplied or added, the way in which they are grouped has no effect on the outcome.

Base

A number that is raised to an exponent.

Base ten

The system of numbers we conventionally use in math, in which each place value represents groups of a different power of ten.

Binomial

An expression of two terms linked together by addition or subtraction.

Brackets

The grouping symbols [and]. Brackets can be used in the same role and purpose as parentheses once parentheses have already been used.

Canceling

Removing a common factor from both the numerator and denominator of a fraction, which does not change the overall value of the fraction.

Common denominator

A number that is a multiple of all the denominators in the problem. Each denominator can be converted by multiplying each denominator by an integer.

Common factors

Integers that are factors shared by two or more integers.

Common multiples

Integers that are multiples of two or more integers.

Commutative property

The property of numbers that says that when numbers are being multiplied or added, the order in which they are multiplied or added has no effect on the outcome.

Complementary

The description of two angles whose degree measures sum to 90°.

Complex fraction

A fraction that has a fraction or fractions in its numerator, denominator, or both.

Complex polygon

A polygon in which at least two of the sides intersect one another. It is not subject to many of the rules regarding simple polygons.

Composite number

An integer that has more than two distinct positive factors. Positive integers other than the number one are either prime or composite.

Concave polygon

A polygon in which at least one of the interior angles is greater than 180°, making at least one of the angles point inward instead of outward.

Convex polygon

A polygon in which all of the interior angles are less than 180°, making none of the angles point inward.

Counting number

A synonym for *natural number*, a counting number is a positive integer such as one, two, or three.

Cross-multiply

A way of solving an equation in which two fractions are set equal to one another by multiplying the numerator of each fraction by the denominator of the other fraction and setting the resulting two values equal to one another.

Cube

The result of a number being multiplied by itself three times.

Cube root

A number that produces a given number when cubed.

Decimal point

The point between the units digit and the tenths digit that divides the number of units from the portion of a unit in a number.

Degree
The unit of measure for an angle. A line has a degree measure of 180°.

Denominator
The bottom number of a fraction.

Density property
The property that defines the truth that there is always another real number between any two real numbers on the number line.

Difference
The result of subtracting one number from another.

Distributive property
The property of numbers that explains that one number multiplied by each term in a polynomial in parentheses yields the same result as multiplying that number by the value of the polynomial.

Dividend
In a division problem, the number being divided by another number.

Divisible
A quality of a number being able to be divided evenly by another number, yielding an integer result.

Divisor
In a division problem, the number dividing another number.

Equation
A math problem that involves an equal sign with one or more numbers, variables, symbols, and operators on both sides of the equal sign.

Equilateral triangle
A triangle in which all sides are of equal length and all angles measure sixty degrees.

Estimate

To round a number and represent it as something to which it is approximately equal, or to use the relative size of numbers in a problem to approximate a value.

Even number

An integer that is divisible evenly into two groups because it has the number two as a factor.

Expanded notation

A way of writing a number by representing each place value multiplied by the size of each group represented by that place value.

Exponent

A number written in small notation raised to the right side of the base number that tells how many times the base should be multiplied as a factor.

Expression

A representation of a value by the use of numbers, variables, and operators, but without an equal sign or inequality.

Factor (noun)

An integer by which another number is divisible.

Factor (verb)

To break a number down into its factors.

Factor tree

The visual representation of factoring an integer until all its prime factors are listed.

Formula

A relationship or rule expressed in terms of math using at least one variable.

Fraction
A number written as a numerator over a denominator in order to represent the value of the numerator divided by the denominator.

Function
A mathematical relationship in which each input to an equation or formula yields exactly one output value.

Greater than
The inequality symbol >, which shows that the value on the left side of the inequality is greater in size than the value on the right side of the inequality symbol.

Greater than or equal to
The inequality symbol ≥, which shows that the value on the left side of the inequality is greater or equal in size than the value on the right side of the inequality symbol.

Greatest common factor
The largest factor that two or more numbers have in common.

Hundreds
The second decimal place to the left of the units digit whose value indicates how many groups of one hundred contribute to the value of the number.

Hundredths
The second decimal place to the right of the decimal point, whose value indicates how many groups of one-hundredth contribute to the value of the number.

Hypotenuse
The name for the longest side of a right triangle, which is the side opposite the right angle.

Identity property of addition

The property stating that adding zero to any number does not change its value.

Identity property of division

The property stating that dividing any number by one does not change its value.

Identity property of multiplication

The property stating that multiplying any number by one does not change its value.

Identity property of subtraction

The property stating that subtracting zero from any number does not change its value.

Improper fraction

A fraction where the numerator is greater than the denominator, causing it to have a value greater than one.

Inequality

A representation that shows the size relationship between two or more values by indicating via symbol that one side is greater than or less than the other.

Integer

A number that can be written without a fraction or decimal, including positive and negative whole numbers and the number zero.

Invert

To flip something over; to take the reciprocal of a number.

Irrational number

A number that is not rational, meaning that it cannot be written as a fraction of a natural number over an integer.

Irregular polygon

A polygon where all sides are not of equal length and all angles are not of equal measure.

Isosceles right triangle

A right triangle with two legs of equal length and two angles, across from those sides, measuring 45° each.

Isosceles triangle

A triangle with two sides of equal length and two angles, across from those sides, of equal degree measure.

Least common multiple

The smallest number that is a multiple of the two or more numbers in question.

Legs

The sides of a right triangle that form the right angle; the two sides that are not the hypotenuse.

Less than

The inequality symbol $<$, which shows that the value on the left side of the inequality is smaller in size than the value on the right side of the inequality symbol.

Less than or equal to

The inequality symbol $\leq$, which shows that the value on the right side of the inequality is greater or equal in size than the value on the left side of the inequality symbol.

Line

A straight geometrical object that exists in one dimension, being both infinitely thin (having no width) and infinitely long (extending forever in both directions).

Line segment
A part of a line that has an end point on both ends. It has a length that does not go on infinitely in either direction.

Lowest common denominator
The smallest common denominator of the two or more numbers in question.

Magnitude
A synonym for *absolute value*.

Mean
The average value of a list of numbers, found by adding the value of all the numbers in the list together and dividing by the number of terms in the list.

Median
The middle number in a list of numbers when the numbers are listed in order from least two smallest. If there are two middle numbers, their mean is the median.

Mixed number
A representation of an irregular fraction as an integer and a proper fraction.

Mode
The number that appears most often in a list of numbers. If two or more numbers appear equally often, they can each be modes.

Monomial
An expression with only one term.

Multiple
A number that is formed by multiplying the initial value by any integer. Each integer has an infinite number of multiples.

Multiplicative inverse
The reciprocal of a number; the number by which an initial number can be multiplied to yield a value of one.

Natural number
A counting number; a positive integer.

Negative
A number with a value less than zero.

Number line
A visual representation of numbers in order of size, with numbers placed further to the right the bigger their size and further to the left the smaller their size.

Numerator
The top number of a fraction.

Obtuse
An angle whose measure is greater than 90°.

Odd number
An integer that cannot be evenly divided into two groups because it does not have the number two as a factor.

Operation
A process that can be performed on two or more numbers, such as addition, division, multiplication, or subtraction.

Order of operations
The system of rules that demands that equations be solved and expressions be simplified in the order of parentheses, exponents and roots, multiplication and division, and addition and subtraction.

Parallel
A description of lines or line segments that don't intersect (or would never intersect if they were extended to infinity) because they are or would be at the same angle in relationship to any other line or line segment.

Parallelogram
A quadrilateral with two sets of parallel sides.

Parentheses
The symbols (and), used to group operations together or to represent multiplication when placed next to one another.

PEMDAS
An acronym used to help remember the order of operations, which demands that equations be solved and expressions be simplified in the order of parentheses, exponents and roots, multiplication and division, and addition and subtraction.

Percent
A way of representing a part of the whole as the numerator of a fraction when the denominator is equal to one hundred.

Percentage
Another word for *percent*.

Perfect square
An integer whose square root is an integer.

Perimeter
The measure of the length of the distance around a two-dimensional shape; the sum of the length of all the sides of a two-dimensional shape.

Perpendicular
A description of two lines or line segments that intersect at a ninety degree angle.

Place value
The value given by a digit based on the position of that digit in the number, with each place value being ten times greater than the place value immediately to its right.

Polygon

A two-dimensional closed shape made up of three or more line segment sides.

Polynomial

An expression in which two or more terms are combined by addition or subtraction.

Positive

A number with a value greater than zero.

Power

An exponent.

Prime factorization

A complete list of the prime factors that make up a number.

Prime notation

A way of writing a number so that it's represented by each of its prime factors raised to the appropriate exponent.

Prime number

An integer that has exactly two distinct positive factors: one and itself.

Product

The result when two or more numbers are multiplied by one another.

Proper fraction

A fraction in which the denominator is bigger than the numerator.

Proportion

A fraction.

Quadrilateral

A polygon with four sides.

Quotient
The result of dividing one number by another.

Radical
A number represented with a root symbol, such as a square root.

Range
The difference between the biggest number and the smallest number in a given list of numbers.

Ratio
A fraction that shows the comparison or relationship between two values.

Rational number
Any number that can be represented as an integer or fraction with a numerator that is a natural number and a denominator that is an integer.

Real number
Any rational or irrational number.

Reciprocal
The result of exchanging the numerator and denominator of a fraction; the value that yields a product of one when multiplied by a number.

Rectangle
A quadrilateral in which all angles measure 90° and each pair of opposite sides are of equal length.

Reduce
To simplify a fraction by factoring a common factor out of both the numerator and the denominator, preserving the initial value of the overall fraction but making the numerator and denominator each smaller in direct proportion.

Regular polygon
A polygon in which all sides are of equal length and all angles are of equal degree measure.

Rhombus
A quadrilateral in which all four sides are of equal length and each pair of opposite sides is parallel.

Right triangle
A triangle in which one of the angles is a right angle, measuring 90°.

Round
To approximate the value of a number by representing it as the number to which it is closest, as determined by a particular place value.

Simple polygon
A polygon that has only one boundary, so that none of the boundary lines intersect with one another between vertices.

Simplify
To combine like terms.

Size
The value of a number; the farther to the right a number is on the number line, the greater its size.

Solve
To determine the possible values.

Square
A quadrilateral in which all sides are of equal length and all angles measure 90°. Also, the result when two of the same number are multiplied by one another.

Square root
A number that produces a specific amount when it is multiplied by itself.

Substitute
To plug in a value for a variable that is or could be equal to the value of that variable.

Sum
The result of adding two or more numbers together.

Supplementary
The description of two angles whose degree measures sum to 180°.

Tens
The first decimal place to the left of the units digit, whose value indicates how many groups of ten contribute to the value of the number.

Tenths
The first decimal place to the right of the decimal point, whose value indicates how many groups of one-tenth contribute to the value of the number.

Terms
Combinations of one or more numbers and variables that do not include addition or subtraction symbols unless they are contained in parentheses.

Trapezoid
A polygon in which two sides are parallel to one another.

Triangle
A polygon with three sides.

Two-dimensional
The description of a flat shape that can be drawn in the two-dimensional plane, such as a triangle or a square.

Units
The ones place value of a number, directly to the left of the decimal point.

Variable
A letter used to hold the place of a number in an equation or expression.

Vertex

The corner of a shape or intersection of two lines or line segments. The plural of *vertex* is *vertices*.

Whole number

An integer; a number that can be written without a fraction or decimal.

Zero

An even integer with an absolute value of zero, indicating no value. Numbers cannot be divided by zero.

Index

Note: Page numbers in *italics* indicate solutions to exercises.

F

G

H

I

L

M

N